THE COMPLETE GUIDE TO AI IN EDUCATION

Harness Artificial Intelligence to Enhance Student Engagement, Build AI
Literacy, Face Ethical Challenges, and Achieve Greater Classroom
Efficiency

GLORIA LEMBO

CONTENTS

INTRODUCTION

Picture a classroom alive with energy—students enthusiastically leaning in, asking questions, and exploring new ideas. Each inquiry sparks a thoughtful response, and every challenge is met with support tailored to their needs. Behind the scenes, artificial intelligence is seamlessly enhancing the learning experience, adapting to each student in real time. This is happening now, and your classroom could be next.

AI in education is not just a tool but a transformative force that empowers both students and teachers. As technology continues to evolve at an astonishing pace, educators must be prepared to integrate AI into their classrooms. While headlines focus on AI's economic impact, its real power lies in its ability to make learning more engaging, accessible, and effective. AI isn't about replacing teachers; it's about amplifying their ability to connect with students, personalize instruction, and create meaningful learning experiences.

But AI also comes with challenges. Students are already using AI to complete their homework, raising concerns about academic integrity and genuine learning. How do we ensure AI is a tool for learning rather than a shortcut to avoid it? At the same time, many students are falling behind due to learning gaps, and educators are asking: Can AI help struggling students catch up, or does it widen the divide? These are crucial questions, and this book will explore how AI can be leveraged responsibly and equitably to support all learners.

Other concerns, such as privacy, data security, and ethics, also deserve careful consideration. Many educators feel uncertain—wondering if AI is too complex, if it's just another passing trend, or if it could diminish the human connection that makes teaching so powerful. These concerns are real, and this book will address them head-on. AI is not a magic solution, but when used responsibly and intentionally, it can be a powerful ally in education.

This book is designed to be your guide, breaking down AI into clear, practical steps. We'll explore the fundamentals of AI in education, showcase real-world success stories, and provide actionable tools and strategies to help you integrate AI into your teaching. You'll gain insights from educators who have successfully used AI to enhance student engagement, build AI literacy, and navigate ethical challenges. You'll also read about failures and lessons learned.

By the time you finish this book, you'll have the knowledge and confidence to harness AI effectively in your classroom. You'll understand what AI can do and how to use it in a way that aligns with your values as an educator. Most importantly, you'll be part of a growing movement of educators who are shaping the future of education—one informed, intentional step at a time.

CHAPTER 1: UNDERSTANDING AI IN EDUCATION

You walk into a classroom where every student has a personalized learning plan. Some are tackling math problems at their own speed, while others dive into interactive lessons that adapt to their individual needs. The teacher isn't just lecturing at the front of the room—they're moving around, offering support exactly where it's needed most. This is the potential reality AI brings to education. As educators, you're on the front lines of this transformation. Understanding AI is crucial for staying relevant and opening up new possibilities for your students. This chapter will guide you through the basics of AI, explain the jargon, and show you how these concepts apply directly to your classroom.

WHAT IS ARTIFICIAL INTELLIGENCE? AI BASICS FOR EDUCATORS

In simple terms, artificial intelligence is the science of making machines smart. Unlike traditional software that follows explicit instructions, AI systems continually learn from data and experiences, adapting and improving over time without constant human intervention. A practical example is the AI that powers your smartphone's voice assistant. It learns how you speak and refines its responses based on your interactions. In education, AI can take many forms, from algorithms that grade essays to systems that suggest personalized learning activities. The key difference between AI and traditional software is AI's ability to learn and adapt, making it a more powerful and flexible tool.

Think of traditional software like building a LEGO car from an instruction manual. You have a clear set of steps, specific bricks, and a precise outcome in mind. Every brick (or line of software code) has a defined place and function. If you follow the instructions exactly, you'll get the same car every time. The rules are fixed, and the program (or the LEGO car) will only do what it was explicitly designed to do. If you want to build something different, you need a new set of instructions.

Now, think of AI as giving the LEGO set to someone without specific instructions but telling them, *"Build me something that moves fast."* Instead of following rigid steps, they experiment with different designs, learn from trial and error, and eventually come up with something that achieves the goal of moving fast. AI systems learn from data (just like experimenting with different LEGO designs), can handle uncertainty, learn from new input, and improve over time. The more they practice and iterate, the better they get.

In short:

- **Traditional software** is like step-by-step LEGO instructions for a predictable fixed result.
- **AI** is like giving LEGO pieces and a goal, letting the system learn and figure out the best design.

To help you feel more at home with AI discussions and enable you to engage with more confidence, the following recaps some of the essential AI terminology you should know.

- **AI:** Artificial intelligence, or AI, is a field of computer science focused on creating machines to perform tasks that would normally require human intelligence. The term "AI" serves as an umbrella, covering all types of programming that enable machines to mimic human cognitive abilities such as learning, reasoning, and problem-solving. By analyzing vast amounts of data and using algorithms to identify patterns, make decisions, or predict outcomes, AI systems have the remarkable ability to learn and improve over time. This adaptability allows AI to tackle a wide range of complex tasks, from translating languages and recognizing images to powering autonomous vehicles.
- **Algorithm:** Algorithm is a term that sits at the heart of AI and plays a crucial role in how machines perform specific tasks. At its simplest, an algorithm is a set of instructions guiding a system step-by-step toward a desired outcome. For instance, in education, adaptive learning platforms like Khan Academy use AI algorithms to tailor content to each student's unique needs. If a student struggles with a particular math concept, the AI

algorithm identifies this and adjusts the material by providing simpler problems or additional resources for better understanding. Similarly, your smartphone employs AI algorithms to determine the order of posts in your social media feed based on past interactions, showing you content you're more likely to enjoy. This ability to learn from behavior and adapt accordingly highlights AI's transformative and dynamic nature in education and everyday life.

- **Data:** Data is the fuel that powers the learning algorithms. Datasets can include anything from text, images, and audio to numerical data, labeled or categorized to provide context for the AI to learn from. In an educational context, this could be anything from test scores to student behavior analytics. The quality and diversity of a dataset are crucial, as they directly influence the accuracy and effectiveness of the AI's performance.

- **Machine Learning**: Machine learning is a subset of AI that teaches computers to learn from data and improve their performance over time without being explicitly programmed for every task. For instance, email platforms use machine learning to filter out spam. They analyze incoming messages and learn to distinguish between spam and legitimate mail based on various indicators—such as sender information, keywords, and user interactions with similar emails. This continuous learning process allows the system to adapt and effectively filter spam as tactics evolve, all thanks to machine learning.

- **Deep Learning:** Deep learning goes beyond traditional machine learning, serving as a specialized subset that uses **multi-layered neural networks** to analyze large volumes of data, identify patterns, and make decisions. These neural networks are designed to mimic the way the human brain processes information by passing data through interconnected nodes, where it is progressively refined to deliver a result. For example, Netflix employs massive data banks to bring you the best user experience each time you open the platform. Deep learning algorithms process user preferences, viewing habits, and even minute interactions to recommend movies and TV shows with astonishing accuracy. This capability is not just limited to entertainment; deep learning is also helpful in medical diagnosis, where it can help detect anomalies in medical images, potentially leading to early detection of diseases such as cancer. In education, deep learning helps develop tools that use image and speech recognition, which can be invaluable for language learning. For instance, a student could use a speech recognition app to practice a new language, receiving instant feedback on pronunciation.

- **Generative AI:** Generative AI is becoming more and more popular thanks to its focus on creating new content rather than simply analyzing data. This branch of AI involves models that generate original text, images, music, and complex simulations. For example, ChatGPT, a natural language model, can produce new and original text based on a few user-entered prompts, aiding in tasks from drafting emails to creating detailed reports. Similarly, DALL-E, another AI model, is used to create realistic images that range from realistic scenes to surreal and imaginative artwork, showcasing the potential of AI in creative fields such as art, design, and visual storytelling.

- **Large Language Model**: Large language model (**LLM**) is a type of artificial intelligence designed to process and generate human-like text. These models are trained on vast amounts of text data from various sources, enabling them to understand and produce language that closely mimics human communication. Examples in business are writing reports, automating customer support, and drafting professional documents. In healthcare, it assists with patient communications and generating medical reports. In education, LLMs provide personalized tutoring, grading, and creating tailored lesson plans.

- **Natural Language Processing:** Natural language processing (**NLP**) is an AI technique that enables computers to understand and process human language. An everyday example of NLP is the translation features in apps like Google Translate. Whether trying to decipher a menu in a foreign language or translating a webpage, NLP algorithms analyze the text and convert it into your native language while maintaining as much of the original meaning as possible. These algorithms are required to understand the complexities of human language, from varying sentence structures to idioms and slang. This serves to highlight the sophisticated capabilities of NLP in bridging language barriers.

Incorporating AI into your teaching practice involves understanding these terms and technologies and recognizing their potential to support your work. By understanding the basics of AI, you're not only enhancing your teaching toolkit but also preparing your students for a world where AI is increasingly prevalent.

Many low-cost and free resources are available online for those eager to explore these concepts further. Platforms like **Coursera** and **Khan Academy** offer courses specifically tailored to beginners, covering everything from the basics of algorithms to more advanced topics in machine learning and neural networks. Engaging with these resources can deepen your understanding and appreciation of AI while empowering you to participate more actively in conversations about how this technology shapes our world.

THE EVOLUTION OF AI IN EDUCATION

The concept of using machines to aid teaching isn't new. In the 1960s, educational technologists began exploring how computers could support learning. Early experiments included simple tutoring systems, which were rudimentary by today's standards but groundbreaking at the time. They hinted at the potential for technology to transform education. Fast forward to the 1980s and 1990s, when AI started gaining traction with the development of expert systems. These programs could mimic the decision-making abilities of a human expert and were used in some educational contexts to offer tailored student support.

As we moved into the 21st century, AI technologies began to accelerate. The internet's rise and the boom in data availability provided fertile ground for AI to grow. During this time, machine learning emerged as a powerful tool, capable of analyzing vast amounts of educational data to uncover patterns and insights. This era saw the introduction of intelligent tutoring systems that could adapt to individual learning styles, not just offering static lessons but evolving based on student interaction. The landmark moment came when schools began integrating AI to support teachers and enhance the student learning experience. This shift marked the beginning of AI's significant presence in educational settings.

One of the key breakthroughs in AI education technology has been the development of **Adaptive Learning Platforms**. These systems assess a student's strengths and weaknesses, creating a personalized learning path that adjusts in real time. Companies like Khan Academy have employed AI algorithms to tailor learning experiences, improving student engagement and outcomes. This approach has revolutionized how students interact with educational content, making learning more personalized than ever before. Another leap forward has been the use of AI-driven assessment tools. These systems go beyond traditional testing methods, analyzing student responses to provide detailed feedback that helps teachers pinpoint areas for improvement.

Predictive Analytics represents another milestone in AI's educational evolution. Schools now use AI to predict student performance, identifying those at risk of falling behind. This allows educators to intervene early, offering support to students before they encounter significant challenges. For example, in New York City schools, predictive analytics is applied through programs like "Teach to One," which customizes math education for middle school students. By analyzing performance data, the program predicts the most effective instructional methods for individual learners, enhancing their educational experience.

Virtual Reality (VR) powered by AI is one of the most exciting recent advancements. It offers immersive learning experiences that transport students to different times and places without leaving the classroom. Picture an immersive history lesson where students experience ancient civilizations or a science class where they dive into the human body. AI creates these dynamic environments, adapting scenarios based on student interaction to ensure optimal learning. This technology not only captivates students but also aids in comprehending complex subjects.

Early adopters of AI in education faced significant challenges, from the high cost of technology and data privacy concerns to resistance from educators unfamiliar with these systems. However, their pioneering efforts paved the way for widespread adoption. Today, AI is more integrated into education than ever, supporting everything from curriculum development to classroom management.

The journey of AI in education is far from over. As technology continues to evolve, so too will its applications in classrooms. Today's educators have a unique opportunity to embrace these tools, enhancing their teaching and improving student outcomes. By understanding the evolution of AI, educators can better appreciate its current capabilities and potential future applications.

THE ROLE OF AI IN MODERNIZING EDUCATION

AI is already driving a profound shift in how education is delivered, redefining the teacher's role from a primary knowledge provider to a facilitator of an interactive and dynamic learning environment where technology becomes a trusted partner. This is the transformative potential AI brings to education. It's not about replacing traditional teaching methods but enhancing them—making education more efficient, personalized, and engaging.

One of AI's most impactful contributions is its ability to personalize learning. AI creates tailored learning pathways for each student by analyzing their data, enabling them to progress at their own pace. This personalization ensures that students are neither overwhelmed by material that is too challenging nor disengaged by content that is too easy. Essentially, AI functions like a personal tutor for every student—a long-standing aspiration of traditional education that is now becoming a reality. This approach leads to better engagement and improved learning outcomes, unlocking the potential of every learner.

AI also significantly enhances classroom efficiency. Time-consuming administrative tasks, such as grading and attendance tracking, can now be automated. This shift allows teachers to focus more on instruction and meaningful student interactions. By freeing educators from routine tasks, AI enables them to dedicate more energy to fostering creativity, critical thinking, and deeper connections with their students. The result is a streamlined educational experience that benefits both teachers and learners, enhancing the quality of teaching and the classroom environment.

Beyond personalization and efficiency, AI empowers educators to create dynamic and responsive curricula. Using real-time data analytics, AI helps teachers adapt lesson plans to effectively meet students' needs. This ensures that learning materials remain relevant, engaging, and impactful. This adaptability is crucial for preparing students for future challenges in an evolving world. With AI, curriculum adjustments are not only swift but also informed by data, ensuring better outcomes for all.

However, integrating AI into education comes with challenges. Some educators may feel hesitant or unsure about incorporating new technologies, fearing that AI could overshadow the human aspects of teaching that are critical for student development. The key lies in balance—leveraging AI to enhance, not replace, the human touch. Teachers remain central to the learning process, guiding and mentoring students in ways that technology cannot replicate.

To overcome these challenges, educators need robust training and ongoing support. Professional development programs focused on AI can empower teachers to confidently integrate these tools into their classrooms. Creating a culture of continuous learning for educators will bridge the gap between traditional teaching practices and modern technological advancements. Schools prioritizing teacher training in AI will see its benefits unfold more quickly as educators embrace innovative approaches and discover creative ways to enhance the learning experience.

Integrating AI in education marks a significant step forward in teaching and learning. It offers the promise of personalized learning, improved efficiency, and adaptable curricula while upholding the vital role of teachers.

DISPELLING MYTHS: WHAT AI CAN AND CANNOT DO

Let's set the record straight. AI isn't a magical solution that replaces educators or turns classrooms into robotic environments. One of the most common misconceptions is that AI will make human teachers obsolete. This simply isn't true. Instead, AI is a tool—an advanced one, but still a tool. As a tool, it can enhance the educational experience by taking over repetitive tasks and offering insights that were previously time-consuming to gather. Think of AI as a trusty assistant, one that can handle mundane duties like grading quizzes or managing schedules, allowing you to focus more on engaging with students, crafting creative lesson plans, and fostering a dynamic learning environment. In this way, AI supports the educator's role rather than overshadowing it.

While AI can analyze data faster and more accurately than any human could, it's important to remember where AI falls short. For instance, AI lacks the emotional intelligence teachers bring to the classroom. It can't read a student's body language to determine if they're confused or disengaged. It can't offer a comforting word or a motivational pep talk when a student is feeling down. These are the nuances of human interaction that AI simply can't replicate. Emotional intelligence, empathy, and the ability to inspire— these are uniquely human traits that are crucial in education. AI can provide data-driven insights, but it takes a teacher to interpret those insights and apply them in a way that resonates with students.

Another myth is that AI can completely personalize education without human input. While AI systems can suggest learning paths based on a student's performance data, they still require oversight. Teachers are needed to adjust these paths based on factors the AI might not account for, like a student's personal interests or emotional state. AI's recommendations, while valuable, are not infallible. They need the human touch to ensure they're aligned with each student's unique needs and circumstances. As such, AI complements your expertise, offering a starting point for deeper engagement rather than a final solution.

Misunderstandings also extend to AI's capabilities in maintaining student motivation. AI can help keep students engaged by providing instant feedback and interactive learning experiences. It can offer challenges tailored to each student's level, helping to maintain their interest and encouraging them to strive for improvement. However, the inspiration and encouragement that drive students to push beyond their limits often come from the relationships they build with their teachers. AI can deliver content but can't replicate the mentor-student bond that motivates students to persevere.

It's also crucial to dispel the myth that AI is flawless. AI systems operate on algorithms and data; if the data is biased, the outcomes will be too. Educators play a key role in monitoring these systems, ensuring that the data used is fair and representative. This vigilance helps prevent AI from perpetuating existing biases and inequalities within education. Teachers are not just users of AI; they are its stewards, guiding its application to ensure it serves the best interests of all students.

In practical terms, AI can dramatically reduce the workload on routine tasks, such as attendance tracking or sorting through student submissions. This automation frees up valuable time, letting teachers dive deeper into lesson planning and student interaction. AI can also help identify trends in student performance, providing early alerts for students who may need extra help. However, the interpretation of this data and the subsequent actions remain firmly in the hands of educators. AI can highlight areas of concern, but it's the teacher's experience and intuition that determine how best to address them.

In essence, AI can significantly enhance the educational environment by performing tasks that would otherwise monopolize a teacher's time, thus allowing for more meaningful interactions between teachers and students. But remember, AI isn't replacing the essential role of educators. Instead, it's a resource designed to enrich the educational process, empowering teachers to do what they do best: inspire and educate the next generation. As we explore further, it becomes clear that the true potential of AI in education lies not in what it can do alone but in what it can help educators achieve.

REAL-WORLD APPLICATIONS: SUCCESSFUL CASE STUDIES

Consider the success story from New York City schools, where initially AI was blocked in schools. Now, AI has become a crucial part of the educational landscape. In collaboration with Microsoft, New York City developed an AI-powered teaching assistant to provide real-time feedback and personalized student support. Here, AI tools assist teachers by analyzing vast amounts of student data, predicting which students might be at risk of falling behind and suggesting timely interventions. This proactive approach helps reduce dropout rates and improve retention. The impact is profound, with teachers reporting not only improved student performance but also a more manageable workload. The key to this success has been strong leadership support and a commitment to ongoing professional development. Teachers receive continuous training, ensuring they're comfortable and confident using AI tools effectively.

In another example, Onion Academy in China has used AI to bridge educational divides. AI platforms deliver high-quality education to students in remote areas, providing tailored learning experiences that meet each student's unique needs. This initiative has been transformative, offering students in underserved regions access to resources they never had before. The success of this program lies in its ability to harness AI's power to democratize education, making quality learning accessible to all, regardless of location.

Special education programs, too, have reaped the benefits of AI. At the CUNY Graduate Disabilities Study program, AI-driven tools create customized learning plans for students with disabilities. These tools help adjust the pace and content of lessons to align with each student's capabilities, leading to a more inclusive and supportive learning environment. The effectiveness of AI in this context is evident in the enhanced learning outcomes and the increased engagement of students who previously struggled to keep pace with traditional methods.

However, not all attempts to integrate AI into educational settings have been smooth sailing. Many schools underestimated the need for comprehensive teacher training, leading to less effective implementations. Without proper guidance and support, teachers often feel overwhelmed or skeptical about adopting new technology. This underscores the critical importance of investing in professional development. Teachers need access to AI tools and the training and resources to confidently integrate them into their teaching practices. When educators are well-equipped with knowledge and skills, they can better harness AI's potential to enhance student learning and engagement.

Several actionable insights emerge from these experiences:

- **Strong Leadership**: School leaders must actively champion AI initiatives. This means providing clear direction, adequate resources, and consistent support to ensure successful implementation. A committed leadership team sets the tone for embracing innovation.
- **Continuous Evaluation**: Regular assessments of AI tools and their impact on student learning are essential. Ongoing evaluation helps identify what's working,

highlights areas for improvement, and ensures the technology remains relevant and effective. Schools should adopt a data-driven approach to refine their strategies.

- **Collaborative Communities**: Building a supportive community among educators can foster a culture of innovation. Teachers need opportunities to collaborate, share their experiences, and learn from one another as they navigate AI integration. Hosting regular workshops, setting up peer-led discussions, and creating mentorship networks of AI-savvy educators can make a significant difference.

Real-world applications of AI in education reveal its transformative potential and highlight the need for thoughtful integration. By learning from successes and missteps, educators can make informed decisions about incorporating AI into their teaching. The path toward AI adoption may have its challenges, but with strong leadership, continuous support, and a commitment to professional growth, schools can unlock its full potential.

CHAPTER 2: IMPLEMENTING AI IN THE CLASSROOM

W hat if you could step into your classroom with an AI assistant by your side, ready to handle routine tasks while you focus on what truly matters—engaging with your students? AI doesn't have to be complicated or intimidating. Starting with simple tools can make a world of difference.

STARTING SMALL: INTRODUCTORY AI TOOLS FOR TEACHERS

As educators, embracing AI can seem intimidating, but it doesn't have to be. The key is to start small and build from there.

Consider tools like **Education Copilot**, an AI teaching assistant that assists with quiz creation, lesson planning, and grading, allowing you to focus on personalized instruction. These applications are perfect for educators new to AI, offering intuitive interfaces and practical functionalities. Free trial versions of such software are often available, providing an excellent opportunity to test the waters without financial commitment. Using these trials, you can explore how AI fits into your teaching style and classroom dynamics, helping you decide on the best tools for your needs.

A successful strategy is to pilot AI tools in a single class or unit to minimize the risk and maximize learning. This approach allows you to tweak your methods and observe direct impacts, making adjustments as necessary. For example, a teacher might start by integrating AI into homework grading. Educators can save hours each week using platforms like **Gradescope**, which employs computer vision algorithms to automate grading. This not only reduces workload but also provides detailed insights into student performance, helping identify areas where students may need additional support.

Generative AI tools, like **ChatGPT**, offer immediate ways to enhance classroom experiences. Using simple prompts, teachers can generate lesson ideas, craft questions, or even simulate class discussions. For instance, you might input a prompt like, *"Generate a list of discussion questions for 'To Kill a Mockingbird,' focusing on themes of justice and morality."* The AI can quickly provide a range of questions, ready to spark thoughtful classroom discussions. This saves time and stimulates creativity, giving teachers a new edge in planning and executing lessons.

Starting small and exploring these introductory AI applications creates a foundation for more advanced integration. As you become more comfortable, you can gradually expand your use of AI, always keeping student engagement and educational outcomes at the forefront. AI doesn't have to replace your teaching methods. Instead, it can enhance and streamline your efforts, providing more time to focus on inspiring and educating your students.

ALIGNING AI WITH CURRICULUM STANDARDS

When integrating AI into your teaching, aligning these tools with curriculum standards isn't just beneficial—it's crucial. Think of AI as a bridge between curriculum goals and student needs. It can support literacy standards by offering personalized reading recommendations based on individual progress. In STEM education, AI applications can simulate complex scientific processes, making abstract concepts tangible and interactive. This alignment ensures that AI tools are not just add-ons but integral parts of your educational strategy, enhancing learning outcomes and meeting educational objectives effectively.

Mapping AI tools to curriculum goals requires a strategic approach. It's about choosing the right tool for the right job. Start by identifying what you want to achieve with your students. Whether improving language arts comprehension or tackling algebraic math equations, AI can support these goals.

- **Literacy:** AI-driven writing assistants and personalized reading platforms can help students meet language arts benchmarks by providing real-time feedback and tailored reading recommendations.
- **Mathematics:** Adaptive AI platforms can adjust problem difficulty in real time, ensuring students grasp foundational skills before advancing.
- **Science:** AI-powered simulations allow students to conduct virtual experiments, reinforcing scientific principles aligned with standardized testing and curriculum guidelines.
- **Assessment & Progress Tracking:** AI-driven analytics can track student performance and provide real-time

insights, helping educators tailor instruction to meet proficiency standards. AI-driven platforms like **Quizizz** can transform assessments into interactive games, making learning more engaging while aligning with educational standards.

Collaboration among educators also plays a critical role in effectively integrating AI into curriculum standards. Working together allows teachers to share insights, exchange best practices, and develop AI-supported lesson plans that align with national and regional learning objectives.

- **AI tools can facilitate teacher collaboration** by streamlining workflows in shared Learning Management Systems (LMS) like **Google Classroom**. Features such as automated grading, personalized learning dashboards, and AI-generated progress reports help educators track student growth and adjust their instruction accordingly.
- **Customizable AI Learning Modules:** Schools can work with AI developers to ensure that educational AI tools align with curriculum needs, allowing for modifications that fit specific school, state, or regional standards. This includes ensuring compliance with regulations protecting data and student information.

Successfully implementing AI in education isn't just about choosing the right apps—it's about setting educators and students up for success.

- **Pilot Testing & Gradual Implementation:** Introduce AI tools in a controlled, low-pressure environment to evaluate effectiveness. Gather feedback from students and teachers to refine AI integration strategies.
- **Teacher Training & Professional Development:** Educators should be equipped with hands-on training to confidently integrate AI into lesson planning. Workshops and professional development programs can help teachers learn how to effectively align AI tools with curriculum goals.
- **Ensuring AI Enhances (Not Replaces) Teaching:** AI should be used as a supplement to, not a replacement for, traditional instruction. Tools should support teachers in delivering high-quality education while maintaining human-centered learning experiences.

By taking these thoughtful steps, educators can ensure that AI isn't just another layer of technology but a powerful, curriculum-aligned tool that enhances both teaching and learning. Through strategic planning, collaboration, and continuous adaptation, AI can become an integral part of the classroom, making learning more personalized, interactive, and effective.

EMBRACING FLIPPED CLASSROOMS WITH AI

Visualize walking into your classroom and finding your students already familiar with the day's material, ready to dive into deeper discussions and hands-on activities. This is the essence of a flipped classroom, and AI takes it to the next level by transforming how content is delivered and how students engage with it. Instead of passively watching videos, memorizing facts, or skimming through materials, students interact with personalized content that adapts to their learning needs. Picture a video lesson that pauses to ask a student a question, offers hints if they're stuck, or suggests extra resources if they're curious to learn more. It's like giving each student a personal tutor to guide them through the material at their own pace so when they walk into class, they're not just prepared but genuinely excited to build on what they've already learned.

One of the biggest advantages of using AI in a flipped classroom is how it supports students before they even step through the door. An AI tutoring system can break down tricky concepts into bite-sized, easy-to-understand lessons, adjusting to each student's pace and learning style. If a student struggles with a math concept, the AI provides extra practice or explains it in a different way—just like a patient tutor. By the time students come to class, they're not just familiar with the material—they're confident and curious, ready to jump into discussions and group activities. AI can even help teachers by suggesting discussion questions or interactive activities based on how students performed during their pre-class work. And once the lesson starts, real-time feed-back tools give teachers instant insights into what's clicking and what needs a little more attention. It's like having an extra pair of hands to make learning more engaging and meaningful for everyone involved.

AI also plays a pivotal role in assessment within flipped class-rooms. Pre-class assessments powered by AI help teachers gauge student understanding of the material, allowing them to tailor in-class activities to address specific knowledge gaps. During and after class, AI tools can evaluate student progress, highlighting concepts mastered and those needing more attention. This continuous feedback loop ensures that every student stays engaged and no one falls behind.

With AI as a partner, the flipped classroom becomes a vibrant, student-centered learning environment. Students actively participate in their education, mastering concepts with confidence and curiosity, while teachers are empowered to provide targeted support and foster a deeper level of engagement.

ADAPTIVE LEARNING TECHNIQUES: MEETING STUDENTS WHERE THEY ARE

Adaptive learning powered by AI provides a classroom where students learn at their own pace, with lessons tailored to their individual needs. By analyzing student performance in real time, AI can adjust the difficulty and style of content, ensuring each learner stays engaged and challenged. This contrasts with static learning paths, which offer the same material to everyone regardless of their unique abilities. Instead, adaptive learning creates dynamic pathways that evolve based on how students interact with the material, providing a customized educational experience that traditional methods just can't match.

The benefits of adaptive learning are especially significant for diverse learners. For students with learning disabilities, AI can provide targeted support, identifying specific areas where they may struggle and offering tailored resources to help them succeed. Advanced students, on the other hand, can benefit from accelerated learning paths that keep them engaged and motivated. A key feature of AI in this context is its ability to conduct diagnostic assessments, pinpointing gaps in understanding with precision. These insights enable educators to intervene effectively, addressing challenges before they become significant roadblocks.

One of the most transformative aspects of AI in education is its ability to identify potential learning disabilities early on. By examining patterns in student performance, AI can flag anomalies that suggest a learning challenge. Educators can then take proactive steps to arrange further assessments and interventions. Beyond identification, AI helps create learning experiences tailored to each student's preferences. Whether a student responds better to visual aids, auditory content, or interactive exercises, AI can adapt the presentation of information accordingly. This customization extends to AI-driven recommendations, guiding students toward further study material that aligns with their interests and strengths.

Several adaptive learning platforms have emerged to support these diverse needs. Tools like **DreamBox** and **MATHia** offer personalized math programs that adjust in real time to each student's level of understanding. **Edmentum** offers adaptive curriculum and assessments to meet individual student needs, supporting personalized learning paths across various subjects.

These platforms often integrate features such as visual aids, audio content, and even language translation, making them ideal for multilingual and inclusive classrooms.

By leveraging these technologies, educators can create an inclusive environment where every student can thrive, regardless of their starting point.

INTERACTIVE AI: FROM CHATBOTS TO VIRTUAL TUTORS

Digital assistants are designed to be available to your students 24/7, ready to answer questions or guide them through complex topics. This is the promise of interactive AI, with chatbots and virtual tutors leading the charge. These tools offer immediate assistance, providing students with resources and support whenever needed. Chatbots, for instance, can handle routine inquiries, freeing you to focus on more impactful teaching moments. Virtual tutors, on the other hand, can provide personalized guidance, adapting to each student's pace and learning style. This interaction can be incredibly valuable, especially in large classrooms where individual attention is often difficult.

Interactive AI plays a pivotal role in education by filling gaps that traditional methods sometimes leave open. **IBM Watson Tutor** is an excellent example of interactive technology that can enhance learning experiences. It uses AI to offer personalized feedback and recommendations, guiding students through their learning journey with tailored support.

Khanmigo, developed by Khan Academy, is a virtual tutor offering tailored feedback and guidance. The following two videos available on YouTube are highly recommended for viewing:

- Khanmigo was featured on a 60 Minutes episode, "Meet Khanmigo: The student tutor AI being tested in school districts" (Dec 9, 2024).
- Sal Khan gave an excellent TED talk, "How AI Could Save (Not Destroy) Education" (May 1, 2023).

Virtual tutoring tools offer benefits like 24/7 availability, ensuring that learning doesn't halt when school ends. However, they also present challenges, such as maintaining student engagement and ensuring content relevance. Without the right balance, students may disengage or fail to see the connection between AI interactions and their educational goals.

Integrating interactive AI tools into your teaching requires thoughtful planning and execution. Start by selecting tools that align with your educational objectives and the needs of your students. It's essential to train students on how to use these tools effectively. Clear guidelines and setting expectations will help them make the most of the technology. For example, encourage students to use chatbots for quick questions or clarifications and rely on virtual tutors for deeper exploration of complex topics. Regular check-ins help ensure that students remain engaged and benefit fully from these resources. As you implement these tools, be prepared to adapt your approach based on student feedback and outcomes, ensuring that the integration of AI enhances rather than disrupts the learning process.

CREATING LESSON PLANS WITH AI SUPPORT

Crafting lesson plans with AI support can revolutionize teaching by structuring lessons that adapt in real time to each student's understanding. AI tools can help you align these plans with your learning objectives, ensuring that each activity and resource serves a clear educational purpose. Start by identifying AI applications that fit your needs. Tools like **MagicSchool AI** are used to assist educators in generating lesson plans, assessments, and differentiated content. Generative AI tools such as **ChatGPT** or **Claude** can be used to create a full lesson plan from scratch using prompts.

Integrating AI into lesson plans means creating interactive lessons that involve students directly in their learning journey. You can incorporate AI components like real-time translation tools in language classes, allowing students to practice, receive immediate feedback, and enhance their language skills. This interactivity keeps students engaged, as they can see the results of their efforts instantly. AI also supports formative assessments by providing instant feedback, which helps students understand their progress and areas where they need improvement. Adaptive quizzes are another excellent tool, as they adjust their difficulty based on student performance, ensuring that each learner is challenged appropriately.

Using AI to help in lesson planning has several advantages:

- **Saves time**: Generates structured lesson plans in minutes.
- **Personalization**: Adapts lessons based on student needs and abilities.
- **Engagement**: Provides interactive elements like AI chatbots, animations, and quizzes.
- **Assessment and Feedback**: Creates personalized assessments and offers instant feedback.

When using AI tools like **ChatGPT** for lesson planning, prompts can be your best friend. These prompts can guide the AI in generating useful content tailored to your lesson needs. For example, for lesson planning, you might use prompts similar to those below to inspire innovative teaching methods:

- *"Generate a lesson plan for teaching fractions to 3rd graders."*
- *"Create a list of engaging activities for a history class on Ancient Egypt."*
- *"Suggest ways to incorporate visual aids in a biology lesson on cell structure."*
- *"Develop a group project for students to explore renewable energy sources."*
- *"Create a role-playing game to teach the principles of economics."*

- *"Suggest modifications for a lesson plan to accommodate students with dyslexia."*
- *"Create extension activities for advanced learners in math."*

Incorporating AI into assessments can be just as transformative:

- *"Design an adaptive quiz for a high school geography class."*
- *"Create a formative assessment for understanding the water cycle."*
- *"Generate feedback for a student's essay on climate change."*

Communication with parents can also benefit from AI, with prompts like:

- *"Draft a newsletter update for parents about upcoming science projects."*
- *"Generate tips for parents to support their child's reading at home."*

SAMPLE LESSON PLANS WITH AI SUPPORT

Putting it all together, the following are some sample AI-generated lesson plan outlines generated by ChatGPT:

1. **AI-Generated Lesson Plan for a Science Class**

Topic: The Water Cycle
Grade Level: 5th Grade
AI-Generated Plan:

- **Objective:** Students will understand the stages of the

water cycle (evaporation, condensation, precipitation, collection).

- **Warm-up Activity:** Show an AI-generated animated video of the water cycle using tools like **SORA**.
- **Lesson Outline:**
  - Introduction to key concepts with AI-powered interactive visuals.
  - AI-generated quiz questions to check prior knowledge.
  - Hands-on experiment: AI suggests simple water cycle experiments using household items.
- **Assessment:** AI-generated worksheet with multiple-choice and open-ended questions.

---

2. **AI-Powered Lesson Plan for English Language Arts (ELA)**

Topic: Creative Writing—Building a Story
Grade Level: 8th Grade
AI-Generated Plan:

- **Objective:** Students will develop a short story using AI-generated prompts.
- **Warm-up Activity:** Use an AI chatbot (e.g., ChatGPT) to generate story starters based on different genres.
- **Lesson Outline:**
  - Students interact with AI to build character profiles and settings.
  - AI suggests improvements in sentence structure and vocabulary.

  - Group discussion on AI-generated vs. human-created writing styles.
- **Assessment:** AI-powered writing assistant provides feedback on grammar, coherence, and creativity.

---

3. **AI-Assisted Math Lesson Plan**

Topic*:* Introduction to Fractions
Grade Level: 3rd Grade
AI-Generated Plan:

- **Objective:** Students will learn to identify and compare fractions.
- **Warm-up Activity:** Use AI to create a real-world scenario (e.g., dividing a pizza into slices) to introduce fractions.
- **Lesson Outline:**
  - Interactive AI-generated visual representations of fractions.
  - AI-adaptive practice problems adjusting difficulty based on student responses.
  - Game-based learning using AI-powered educational apps.
- **Assessment:** AI-generated personalized quizzes based on each student's progress.

---

4. **AI-Enhanced History Lesson Plan**

Topic: The American Revolution
Grade Level: 10th Grade
AI-Generated Plan:

- **Objective:** Students will analyze the causes and impact of the American Revolution.
- **Warm-up Activity:** Use AI to generate historical chatbots (e.g., talking to "George Washington" or "Paul Revere").
- **Lesson Outline:**
  - AI-assisted timeline creation of key events.
  - AI-generated debate prompts on perspectives from British vs. American colonists.
  - Virtual AI-generated field trip through a historical simulation.
- **Assessment:** AI provides instant feedback on essay writing and analysis.

---

5. **AI-Driven STEM Lesson Plan**

Topic: Introduction to Coding with Python
Grade Level: High School
AI-Generated Plan:

- **Objective:** Students will understand basic Python syntax and logic.
- **Warm-up Activity:** Use an AI chatbot to explain simple Python concepts.

- **Lesson Outline:**
  - AI generates coding exercises based on student skill levels.
  - AI-powered debugging assistant helps students fix errors in real time.
  - Interactive AI project: Students build a simple chatbot or game using AI coding support.
- **Assessment:** AI evaluates code efficiency and provides improvement suggestions.

BALANCING AI AND HUMAN INTERACTION: FINDING THE SWEET SPOT

In the rush to integrate AI into classrooms, it's crucial to remember that technology should enhance education, not replace the human touch. While AI can tackle repetitive tasks and offer real-time analytics, it can't replicate the warmth and intuition that educators bring to the learning environment. The heart of teaching lies in the relationships you build with your students. These connections foster trust, inspire learning, and help you guide students through their educational experiences. AI should act as a support tool, providing data and insights that allow you to focus on what you do best: teaching and inspiring young minds.

Maintaining human interaction in classrooms enriched by AI is about finding the right balance. Peer discussions, for example, can be facilitated with AI-generated prompts, sparking deeper conversations and critical thinking among students. In these discussions, students learn to articulate their thoughts, listen to others, and develop social skills—areas where AI falls short. Teacher-led sessions complemented by AI data can provide a more comprehensive understanding of student progress. AI might identify trends or highlight areas where a student struggles, but it's your empathy and experience that interpret this data and decide the best course of action. You know when a student needs encouragement or a different approach, something AI can't intuitively gauge.

Despite AI's advantages, its limitations are clear. It lacks the ability to understand the nuanced needs of individual students. AI can analyze data but doesn't feel the anxiety a student might have before a test or the excitement of mastering a new concept. It's in these moments that your role becomes irreplaceable. You bring the empathy, patience, and adaptability necessary to respond to the varied needs of your classroom. This is where human judgment shines, filling the gaps left by AI's analytical capabilities. You tailor your teaching to each student, understanding that education is not a one-size-fits-all endeavor.

Strategies for balancing AI and human interaction involve integrating AI into your teaching toolkit while ensuring that it doesn't overshadow the personal connections you foster. Encourage students to engage with AI tools for independent learning while using class time to discuss their findings and reflections. Facilitate activities that leverage AI insights to inform group projects or debates, ensuring that students practice collaboration and communication skills. By positioning AI as a valuable assistant rather than a replacement, you maintain the integrity of the educational experience, allowing technology to support your mission to educate with compassion and insight.

CHAPTER 3: ENHANCING STUDENT ENGAGEMENT WITH AI

I n today's digital age, students—often called "digital natives" —crave interactive, multimedia-rich content, instant feedback, and gamified experiences that make learning an adventure. AI can transform these preferences into a dynamic learning environment that keeps students engaged and eager to participate.

AI AND STUDENT MOTIVATION

AI has the remarkable ability to analyze student behavior and identify what truly motivates them. Through AI-driven behavioral analysis, educators can uncover what makes each student tick. This might involve recognizing patterns in engagement, such as when a student participates most actively or what types of content capture their attention. AI can then suggest personalized strategies to enhance motivation, like a reward system that acknowledges achievements in a way that resonates with each student. Whether it's a digital badge for completing a difficult task or a personalized message acknowledging their progress, AI ensures that recognition is meaningful and impactful.

One of AI's most engaging features is its ability to create adaptive quizzes that evolve based on student interaction. These quizzes adjust in real time, offering questions that align with each student's current level of understanding. If a student excels, the questions become more challenging, pushing them to extend their knowledge. Conversely, if a student struggles, the AI provides questions that reinforce foundational concepts before moving forward. This adaptive approach keeps students engaged and builds confidence as they see their progress reflected in the challenges they conquer. The immediacy of these adaptations helps maintain a steady flow of motivation and interest.

AI doesn't just support learning; it actively measures engagement through real-time metrics. Dashboards give educators insights into student participation, highlighting who's involved and who might need a nudge. These analytics can track everything from the number of interactions a student has with the content to their responsiveness in discussions. Having this data at your fingertips allows you to make informed decisions about how to engage your students more effectively. You can identify who might benefit from additional encouragement or who might thrive with more challenging tasks, ensuring no student is left unseen.

Feedback is a powerful tool in education, and AI enhances this process by offering instantaneous and personalized responses to student work. Imagine a student receiving immediate feedback on a math problem they've just solved, with suggestions on how to improve or a quick note of praise. Such timely responses can be incredibly motivating, helping students feel seen and supported in their efforts. AI-generated motivational messages can also offer encouragement and celebrate milestones in a student's learning journey. This feedback helps foster a growth mindset, where students learn to see challenges as opportunities to grow and improve.

GAMIFICATION AND AI

What if you can turn a typical lesson into an engaging game where students readily participate, driven by curiosity and the thrill of competition? This is where AI steps in, incorporating game elements into educational activities to capture students' attention. AI can transform learning into an interactive, fun experience by integrating game points, levels, and leaderboards. These elements foster a sense of achievement and encourage students to push beyond their limits. The magic lies in how AI personalizes these gamified experiences. AI can adjust difficulty levels by analyzing student performance, ensuring each student remains challenged yet not overwhelmed. This adaptability enhances motivation and keeps students invested in their learning.

The impact of gamification on student engagement is significant. AI-driven game scenarios can be tailored to curriculum topics, turning assessments into enjoyable challenges. Consider a history lesson where students embark on a virtual quest to uncover ancient artifacts, earning points for their discoveries. Such activities increase participation as students become active learners, eager to explore and interact with the content. According to a study on AI-driven gamification, teachers have observed improved understanding and retention among students engaged in gamified learning environments. This approach boosts motivation and enriches the learning experience, making education a more dynamic and immersive journey.

To boost engagement, think about creating competitive learning environments with AI. Introduce friendly contests where students can compete individually or in teams, earning rewards for their achievements. These competitions can be designed to reinforce key concepts, encouraging students to revisit and master the material. One successful example is the use of AI-powered platforms like **Kahoot! AI**, which offers interactive quizzes and instant feedback. **Minecraft: Education Edition** uses the popular sandbox game Minecraft for interactive learning. Teachers create lesson-based worlds for subjects like science, math, and history. These tools make learning a social experience, fostering collaboration and healthy student competition. By leveraging AI, educators can transform traditional lessons into vibrant, engaging activities that resonate with today's learners.

Implementing AI-driven gamification into classrooms requires careful planning and execution to maximize its benefits. Select the AI tools that align with your subject and educational goals. Evaluate options based on ease of use, compatibility with existing systems, and the ability to monitor and analyze student progress. To ensure successful adoption, provide educators with comprehensive training on gamification techniques, building their confidence to use these tools creatively and efficiently. Encourage teachers to experiment with various strategies, continuously refining their approach through student feedback and measurable outcomes. By integrating AI-driven gamification thoughtfully, educators can create a more engaging and inclusive educational experience.

Striking the right balance between educational content and game elements is essential for effective AI gamification. The primary focus must remain on achieving learning objectives, avoiding the temptation to prioritize entertainment at the expense of educational value. This balance can be achieved by designing game scenarios that align closely with curriculum goals while maintaining a fun and engaging atmosphere. Inclusivity is another critical consideration—games should be accessible to all students, regardless of their abilities or backgrounds. AI can play a pivotal role by offering adaptive features that address diverse learning needs, ensuring every student can participate meaningfully. While technical difficulties or resistance to change may arise, the potential benefits of AI gamification far outweigh the hurdles. With careful planning and execution, educators can leverage the power of AI to inspire, motivate, and engage students like never before.

VIRTUAL REALITY AND AI

Imagine a classroom where history comes alive: students don VR headsets and stroll through the bustling streets of ancient Rome or stand in the shadow of the pyramids as they're being constructed. With AI and virtual reality (VR) working hand in hand, these immersive experiences aren't just a novelty—they're transforming education. Tools like **ClassVR** provide curriculum-aligned VR content that turns traditional lessons into interactive adventures. AI takes these experiences further by introducing responsive avatars that guide students through historical sites, answering questions, providing context, and even quizzing them to reinforce learning. This blend of interactivity and immersion ensures students remain engaged and actively involved in their education.

AI plays a pivotal role in elevating VR beyond a visual spectacle, transforming it into a personalized, interactive journey. For example, AI-driven tools like **Labster VR** provide virtual science labs where students can conduct experiments in biology, chemistry, and physics. By adapting tasks and challenges based on individual student responses, AI ensures that VR experiences are accessible yet suitably challenging, catering to various learning styles. Visual and kinesthetic learners, in particular, benefit from this hands-on approach as abstract concepts become tangible and easier to understand.

In special education, customizable VR environments are a game-changer. Tools like the **Eduverse by Avantis Education** provide secure, adaptable settings that meet students' unique needs. These environments can offer sensory-friendly options or provide guided prompts to help students progress independently. For students with autism, VR simulations can create safe spaces to practice social interactions, fostering confidence and skills that carry over into real-life situations. AI further enhances these experiences by tailoring the content to the specific needs of each student, ensuring inclusivity and accessibility.

Integrating VR and AI in education isn't just about making lessons more engaging—it's about redefining the learning experience. These technologies enable educators to push the boundaries of traditional teaching, creating dynamic and inclusive environments where every student can thrive.

USING AI FOR COLLABORATIVE LEARNING

AI technologies are revolutionizing how students collaborate and work as a team by offering tools that streamline group projects and enhance peer assessments. These platforms allow students to engage in meaningful discussions, share resources, and collectively solve problems more efficiently than ever before. AI fosters a sense of community by supporting peer-to-peer interactions, making discussions richer and more insightful. Students can use AI-driven platforms to communicate, exchange ideas, and critique each other's work in real time, creating a collaborative environment that encourages learning through interaction.

AI-driven collaboration tools, such as **Microsoft Teams with Viva Learning** and **Kahoot! AI** facilitates student teamwork by enabling educators to create assignments requiring students to collaborate, share insights, and develop solutions collectively. These tools offer seamless integration into educational settings, allowing students to engage in group activities that enhance their learning experience. For instance, Microsoft Teams enables students to create shared documents, participate in group chats, and access learning resources, all within a single platform. Kahoot! AI, on the other hand, gamifies learning by allowing students to create and participate in quizzes together, encouraging healthy competition, and ensuring everyone is on the same page. These tools not only make collaboration more accessible but also foster a sense of belonging and engagement among students as they work toward common goals.

To effectively integrate AI tools that support collaborative learning, it's crucial to establish clear roles and responsibilities within student groups. Assigning specific tasks to each group member ensures everyone contributes and engages with the project, fostering accountability and teamwork. Training students to use AI collaboration tools effectively is equally important. Providing tutorials or workshops can help students understand the functionalities of these platforms, enabling them to utilize the tools to their full potential. This training empowers students to navigate the technology confidently, enhancing their collaboration experiences and allowing them to focus on the learning objectives. More importantly, this approach enhances learning and develops critical life skills, such as communication, problem-solving, and teamwork, preparing students for future collaborative endeavors.

ENCOURAGING CREATIVITY WITH AI

AI tools play a transformative role in fostering creativity in education, empowering students to become creators rather than mere consumers of information. Platforms like **Canva for Education** provide students with user-friendly tools to design digital art, create infographics, and craft visuals that showcase their unique perspectives. These tools extend beyond static images to dynamic media, enabling students to create animations, videos, and even 3D models through applications like **Tinkercad** or **Blender**. These AI-supported platforms simplify complex tasks, allowing students to focus on imaginative expression rather than technical limitations.

Storytelling platforms such as **Book Creator** and **Storybird** leverage AI to inspire narrative creation. These tools offer prompts, plot suggestions, and creative guidance that help students craft compelling stories. For example, an AI might suggest an alternate history scenario or provide character-building prompts, encouraging students to dive into imaginative storytelling and explore new narrative structures. The iterative nature of these tools helps refine their skills, building confidence as they develop their ideas further.

Incorporating AI into creative projects requires thoughtful strategies. One effective approach is using AI-generated prompts to stimulate creative writing. These prompts can range from simple story starters to complex scenarios that challenge students to think critically and creatively. For instance, an AI might suggest a world where gravity functions differently, prompting students to explore how this change affects daily life. Such scenarios encourage students to push the boundaries of their imagination. Collaborative art projects using AI overlays can also be powerful. Imagine students working together, each contributing a piece to a larger digital mural, with AI helping to blend elements seamlessly. This collaboration fosters a sense of community and allows students to learn from each other's creative approaches.

AI is also revolutionizing music education in several innovative ways, enhancing teaching and learning experiences. Platforms like **Yousician** provide real-time feedback on students' performances, helping them improve their technique and timing.

AI's ability to encourage innovation isn't limited to the arts. Tools like **Scratch** and **Minecraft: Education Edition** let students engage in creative coding and world-building, developing problem-solving skills in interactive and exploratory ways. AI-driven simulations, such as those offered by **Labster** or **PhET Interactive Simulations**, immerse students in "what-if" scenarios. For example, students might use these tools to model the effects of climate change on various ecosystems or hypothesize solutions to urban planning challenges. These interactive experiences transform abstract concepts into tangible challenges, encouraging inquiry, exploration, and innovative thinking.

Evaluating AI's impact on creativity involves assessing both tangible outcomes and skill development. Student projects often demonstrate increased originality, showcasing unique perspectives and novel approaches. This iterative process builds critical and divergent thinking skills, encouraging students to explore unconventional methods and push boundaries.

Educators can cultivate an environment where innovation flourishes by integrating AI tools thoughtfully into creative endeavors.

Whether through storytelling, digital art, music, or problem-based learning, AI empowers students to explore the full extent of their creative potential, preparing them for a future that values imagination and ingenuity.

SUCCESSFULLY ENGAGING STUDENTS WITH AI

In classrooms across the globe, educators are tapping into AI's potential to engage students in new and exciting ways. Take, for instance, the innovative approach at Nord Anglia International Schools Rotterdam (NAISR) in the Netherlands, where AI-powered language apps have transformed how students learn English. Using these apps, students receive tailored lessons that adapt to their learning pace, making language acquisition fun and effective. The school's success can be attributed to its commitment to integrating technology that complements traditional teaching methods. Teachers were trained extensively, ensuring they understood both the tool's potential and its limitations. This balance of technology and human interaction has led to impressive improvements in language proficiency and student engagement.

Key to these successes is an understanding of how to maintain student interest over the long term. Successful schools often focus on creating a supportive environment where technology is seen as an ally rather than a replacement for traditional education. Students are encouraged to explore AI tools at their own pace, fostering a sense of autonomy and curiosity. This approach increases engagement and empowers students to take charge of their learning. Additionally, regular feedback sessions between students and teachers help refine the integration of AI, ensuring it remains relevant and effective in meeting educational goals.

NAISR correctly states: *"In our modern age, AI is a valuable life skill. Knowing how to use AI, debating the ethics of AI, and learning about AI safety will soon be critical skills for everyday life as well as in the job market."*

However, the road to successful AI integration is not without its hurdles. Many schools encounter technical issues during implementation, such as software glitches or compatibility problems with existing systems. In one case, a school in Canada faced challenges with an AI platform that frequently crashed, disrupting lessons and frustrating both students and teachers. The solution lies in collaborating closely with the software provider to address these issues promptly. By maintaining open lines of communication and being proactive in troubleshooting, the school was able to stabilize the platform and continue leveraging AI to enhance learning.

Another challenge is ensuring that AI tools are inclusive and accessible to all students. In some schools, a lack of resources or technical support can create disparities in access. Overcoming this requires a concerted effort to provide equal access to technology, whether by providing devices or offering training sessions for students and educators. Schools that have successfully navigated these challenges often have strong support networks in place, including partnerships with tech companies and community organizations that help bridge resource gaps.

Drawing from these experiences, there are several actionable recommendations for educators looking to engage students with AI:

- **Invest in professional development and training** to ensure you and your colleagues are comfortable with the technology. This training should cover not only how to use AI tools but also how to integrate them into existing curricula effectively.
- **Establish a feedback loop with students** to continually assess AI's impact on their learning experience. This ongoing dialogue can provide valuable insights into what's working and what might need adjustment.
- **Foster a culture of experimentation and resilience** within your school. Encourage teachers and students to explore new AI tools and approaches, learning from both successes and setbacks.

CHAPTER 4: PRACTICAL AI APPLICATIONS AND CASE STUDIES

F rom personalized learning in special education to bridging resource gaps in rural schools, AI is proving to be a versatile ally in addressing the unique challenges of education. This chapter explores real-world applications of AI through case studies that highlight its successful integration in various contexts, such as urban or rural classrooms, STEM education, language learning, and homeschooling. Through these stories, educators and policymakers will gain insights into how AI breaks down traditional barriers and creates new opportunities to innovate and inspire in education.

AI INTEGRATION IN URBAN CLASSROOMS

Urban classrooms present a unique setting for integrating AI into education. Students have diverse cultural backgrounds and often face the challenge of high student-to-teacher ratios. These dynamics create both hurdles and opportunities. On the one hand, the sheer number of students can make personalized attention difficult. On the other hand, the cultural richness and the vast array of perspectives provide fertile ground for AI tools that cater to diverse learning needs. In these environments, AI can be a tool to bridge gaps, offering personalized learning experiences that traditional methods might struggle to provide.

AI tools help teachers manage large classes by automating administrative tasks, like attendance and grading, allowing educators to focus more on teaching and less on paperwork. AI-driven content also plays a crucial role in enhancing student engagement. For instance, interactive learning platforms use AI to adapt to each student's level, offering tailored content that resonates with their interests and strengths. This personalization keeps students engaged and motivated, improving both participation and learning outcomes.

The impact of AI on urban student populations extends beyond academics. Those schools that have successfully integrated AI report improved academic performance and heightened student engagement. Access to personalized learning resources means students can explore subjects at their own pace, deepening their understanding and curiosity. This empowerment leads to a more inclusive educational experience where students feel seen and supported. Urban schools leveraging AI also often notice an increase in student confidence and a reduction in dropout rates. These positive outcomes reflect AI's potential to transform urban education into a more equitable and effective system.

Several best practices have emerged for urban schools looking to integrate AI successfully. Collaborative planning with community stakeholders is key. Engaging parents, local businesses, and educational bodies in the planning process ensures that AI tools meet the community's specific needs and goals. This collaboration can also help secure resources and support for AI initiatives, making the integration process smoother and more sustainable. Another strategy is leveraging AI for resource optimization. By using AI to analyze data, schools can identify areas where resources are most needed, ensuring that investments in technology yield the greatest impact.

Additionally, continuous professional development for educators is crucial. Teachers need training not just on how to use AI tools but also on how to integrate them into their teaching practices effectively. When done thoughtfully and collaboratively, integrating AI in urban classrooms holds the promise of a more personalized, engaging, and equitable learning experience.

AI INTEGRATION IN URBAN SCHOOLS CASE STUDY: MIAMI-DADE COUNTY PUBLIC SCHOOLS

Miami-Dade County Public Schools (M-DCPS), the fourth-largest school district in the United States, has successfully integrated AI-driven technologies to enhance education, particularly in STEM learning and language acquisition. With over 300,000 students, many of whom come from bilingual or multilingual backgrounds, M-DCPS has embraced AI to personalize instruction, improve student engagement, and support educators in creating more effective learning experiences.

One of the district's most notable AI-driven initiatives is the integration of Carnegie Learning's AI-powered math platform, **MATHia**. This adaptive learning system uses artificial intelligence to track each student's progress, identify areas of struggle, and provide customized learning pathways. The AI-powered tutor delivers real-time feedback, allowing students to master concepts at their own pace while ensuring teachers receive data-driven insights to tailor classroom instruction. This program has significantly improved math proficiency, particularly in underperforming schools, by helping students build strong problem-solving skills.

M-DCPS has also implemented AI-powered language learning platforms such as **Microsoft Translator and AI-powered English as a Second Language (ESL) tools** to support its large English Language Learner (ELL) population. Given that over 70 percent of students speak a language other than English at home, these AI tools help bridge communication gaps by providing real-time translation during classroom discussions and translating educational materials into multiple languages. This ensures that students and their families remain engaged in the learning process, even if English is not their primary language.

M-DCPS has also partnered with **IBM Watson AI** to pilot AI-driven career readiness programs in high schools. Through AI-powered mentorship chatbots and career exploration tools, students can receive personalized guidance on future career paths, college applications, and job readiness. These AI tools analyze students' interests, strengths, and academic records to recommend customized learning tracks, internships, and scholarship opportunities, helping to bridge the gap between education and workforce development.

To further support its diverse student body, M-DCPS has adopted AI-powered early intervention systems that use machine learning to identify students at risk of failing or dropping out. These AI models analyze data points such as attendance, grades, behavioral patterns, and engagement levels to alert educators about students who need additional support. School counselors can then intervene with targeted academic resources, social-emotional learning programs, and mentorship initiatives, leading to improved graduation rates and student retention.

By leveraging AI for adaptive learning, multilingual education, career readiness, and early intervention, Miami-Dade County Public Schools has created a more inclusive, data-driven, and student-focused educational system. The success of these AI-driven programs highlights the potential of artificial intelligence in urban education, ensuring that diverse student populations receive equitable access to personalized learning and future career opportunities.

AI INTEGRATION IN RURAL SCHOOLS

In many rural areas, the promise of incorporating AI into education feels distant due to several significant hurdles. One of the most pressing challenges is limited internet connectivity. Despite the digital age, many rural communities still struggle with unreliable or slow internet, making it difficult for schools to access AI-based educational tools. A lack of funding and resources compounds this connectivity issue, so even when internet access is available, schools often lack the financial means to invest in the necessary hardware and software. These barriers create a digital divide, placing students in rural areas at a disadvantage compared to their urban peers.

However, innovative solutions are emerging to tackle these obstacles head-on. Community partnerships have become a lifeline, enabling rural schools to access technology and resources that would otherwise be out of reach. By collaborating with local businesses and educational organizations, schools can secure funding and technical support to implement AI-driven projects. Another creative approach is the use of mobile AI learning labs. These labs bring technology directly to students, offering hands-on experiences with AI tools without needing permanent infrastructure. They travel from school to school, providing rural students with the opportunity to engage with cutting-edge technology and learn in ways previously unimaginable.

AI's role in bridging educational gaps in rural areas is profound. Personalized learning opportunities allow isolated students to access a wealth of diverse educational content, leveling the playing field with their urban counterparts. For instance, AI can recommend specific learning paths based on a student's progress, ensuring each learner receives the support and challenges they need. This personalized approach enhances academic achievement and fosters a love for learning as students explore subjects in depth and at their own pace. AI also provides access to a global library of resources, enabling students to connect with content and experts from around the world, broadening their horizons and expanding their educational experience.

When integrated thoughtfully, AI helps rural schools overcome geographical and technological barriers, providing students with opportunities to thrive. The potential is there, waiting to be tapped into with the proper support and strategies. Schools that embrace these innovative approaches offer students the tools they need to succeed in an increasingly digital world. While challenges remain, the growing number of success stories demonstrates that with determination and creativity, rural schools can harness AI to create engaging learning environments that rival those in urban settings.

AI INTEGRATION IN RURAL SCHOOLS CASE STUDY: PIEDMONT CITY SCHOOLS, ALABAMA

One of the most compelling examples of AI-driven educational transformation in a rural setting is Piedmont City Schools in Alabama. Located in a small, economically disadvantaged rural district, Piedmont faced significant challenges related to limited resources, teacher shortages, and student engagement. However, through the strategic implementation of AI-powered learning platforms, the district has successfully bridged educational gaps and enhanced student outcomes.

A major success factor in Piedmont's AI-driven transformation has been the adoption of **DreamBox Learning**, an AI-powered adaptive math program that personalizes instruction based on student performance. Given the challenges rural schools often face in recruiting specialized math teachers, DreamBox has helped fill this gap by providing students with real-time, tailored instruction that adapts to their learning speed. As a result, math proficiency scores have significantly improved, and students who previously struggled with foundational concepts are now more confident and engaged.

To support literacy development, Piedmont implemented **Lexia Core5 Reading**, an AI-powered literacy platform that provides individualized reading instruction based on real-time performance data. Many rural students in Piedmont enter school with limited access to early literacy resources, and AI has played a crucial role in identifying struggling readers and adjusting instruction accordingly. Teachers receive AI-generated reports that pinpoint specific skill gaps, allowing them to intervene effectively and ensure students stay on track.

Piedmont City Schools has also leveraged AI-powered early warning systems to track student attendance, engagement, and academic performance. Using machine learning algorithms, the district has been able to predict which students are at risk of falling behind and implement proactive interventions such as one-on-one tutoring, counseling, or mentorship programs. These AI-driven insights have contributed to lower dropout rates and higher student retention, a critical achievement in rural education where students often face economic and social barriers to graduation.

Another key AI integration in Piedmont has been the use of AI-powered virtual tutoring systems like **Edmentum** to compensate for teacher shortages in advanced subjects. Rural districts often struggle to offer specialized courses such as AP math, science, or foreign languages. However, AI-driven virtual tutoring platforms provide students with on-demand access to expert instruction, ensuring they receive high-quality education regardless of location.

Through its commitment to AI-driven education, Piedmont City Schools has transformed a once-struggling rural district into a model for innovation. The integration of adaptive learning, early intervention, and AI-powered tutoring has leveled the playing field for students in an underserved region, proving that AI can be a powerful tool for overcoming the unique challenges of rural education. Piedmont's success inspires other rural districts looking to leverage AI to improve student outcomes, teacher effectiveness, and overall educational equity.

AI IN SPECIAL EDUCATION

AI tools are revolutionizing how educators approach teaching students with special needs, providing tailored solutions that were once unimaginable. These technologies are not just about making tasks easier; they are about transforming educational experiences to be more inclusive and effective.

There are now excellent AI tools specifically geared for special education's unique needs:

- **Developmental Disorders**: For autism and other developmental disorders, AI offers assistive technologies that cater to specific needs. Tools like **MagicSchool.ai**, **Easy-Peasy.AI,** and **EdTech's Expert IEP** help create comprehensive and accessible Individual Education Plans (IEPs), reducing the administrative burden on already stretched-thin teachers. For students, this means receiving educational plans that truly reflect their unique learning paths, enabling them to thrive in environments that understand and accommodate their needs.
- **Speech and Communication Challenges**: For students with speech and communication challenges, AI-driven Augmentative and Alternative Communication (AAC) tools like **Proloquo2Go** and **Tobii Dynavox** offer life-changing support. These platforms use AI-based symbol prediction and eye-tracking technology to enable nonverbal students or those with limited mobility to express themselves. Such tools are essential for students with autism, cerebral palsy, or speech disorders, allowing them to participate more fully in classroom activities and social interactions.

- **Learning Disabilities:** AI also plays a crucial role in literacy support and writing assistance. Students with dyslexia or other learning disabilities often struggle with reading and writing tasks. However, AI-powered tools like **Read&Write by Texthelp**, **Speechify**, and **Co:Writer Universal** provide real-time reading assistance, word prediction, text-to-speech, and speech-to-text functionalities. These applications help students decode text, improve comprehension, and develop stronger writing skills by minimizing frustration and cognitive overload.

- **Hearing or Visual Impairments:** For students with hearing or visual impairments, AI-driven assistive technologies significantly improve accessibility. **Microsoft Immersive Reader** reads text aloud and highlights words, making it easier for students to follow along. Meanwhile, **Seeing AI** and **Google Lookout** describe objects and text in the environment, offering independence and enhanced learning experiences for visually impaired students.

- **Emotional or Behavioral Support:** Beyond academics, AI is being leveraged for emotional and behavioral support in special education. AI-driven applications like **BrainCo** use EEG headbands to monitor students' focus and engagement, helping teachers tailor interventions for students with ADHD. Additionally, mental health chatbots like **Woebot** and **Wysa** offer AI-powered emotional support, helping students manage social anxiety and stress through guided conversations. In autism education, the Autism Glass Project uses AI-powered smart glasses to help students recognize emotions and develop social skills, making interactions more intuitive.

Integrating AI into special education transforms the learning experience by making it more inclusive, accessible, and personalized. These technologies empower students with disabilities to overcome barriers and reach their full potential while also providing educators with valuable tools to enhance their teaching strategies. As AI continues to advance, its role in special education will expand, opening new possibilities for supporting students with diverse learning needs in ways that were once unimaginable.

However, integrating AI into special education is not without its challenges. One major hurdle is the initial resistance to adopting new technologies, often due to a lack of familiarity or fear of technology replacing human interaction. Additionally, there are concerns about data privacy and ensuring that AI systems are free from bias. Addressing these concerns requires a commitment to ethical practices and regular evaluations of AI implementations to ensure they remain effective and fair.

Despite these challenges, the benefits of using AI in special education are profound. Personalized learning is at the heart of this transformation. AI's ability to offer tailored educational experiences means that students who might have been left behind in traditional settings now have the opportunity to excel. By embracing these technologies, educators can create truly inclusive learning environments, giving every student the chance to succeed.

AI IN SPECIAL EDUCATION CASE STUDY: CHICAGO PUBLIC SCHOOLS

One notable example of AI's successful integration in special education is **Chicago Public Schools (CPS)**, which has leveraged AI-driven tools to support students with diverse learning needs across its urban school districts. With over 330,000 students, including thousands requiring special education services, CPS has implemented various AI-powered solutions to enhance accessibility, personalize instruction, and improve student outcomes.

One of the most impactful implementations has been using **Microsoft Immersive Reader** in CPS classrooms to support students with dyslexia and reading difficulties. The tool, embedded within educational platforms like Microsoft Word and OneNote, allows students to have text read aloud, adjust font sizes, and translate content into different languages—an essential feature in a district where many students come from multilingual backgrounds. The AI-driven accessibility features have improved reading comprehension, boosted student confidence, and allowed special education teachers to tailor literacy instruction more effectively.

Additionally, CPS has integrated **Otter.ai**, an AI-powered transcription tool, to assist students with hearing impairments. The tool provides real-time captioning of lectures, ensuring that students can follow along without missing important classroom discussions. This has been particularly beneficial in large, noisy urban classrooms where traditional hearing aids may not fully capture the teacher's voice. Teachers have reported increased engagement from deaf and hard-of-hearing students, who can now participate more actively in lessons and discussions.

Another AI-driven success in CPS has been the **DreamBox Learning** adaptive math program, which uses AI algorithms to personalize instruction based on a student's unique learning needs. In special education classrooms, where students often require individualized attention, DreamBox has helped bridge learning gaps by adjusting the difficulty of problems in real time. Special educators have praised the program for providing data-driven insights that help them identify areas where students are struggling and intervene accordingly.

Beyond academics, CPS has experimented with **AI-powered social-emotional learning (SEL) tools** like **Woebot**, a chatbot designed to support students with anxiety and behavioral challenges. By offering guided conversations, cognitive behavioral therapy techniques, and emotional check-ins, the AI assistant has provided students with an additional layer of mental health support, complementing the work of school counselors.

These AI-driven initiatives in Chicago Public Schools demonstrate the power of artificial intelligence in enhancing special education in an urban setting. By leveraging AI for accessibility, personalized learning, and emotional support, CPS has created a more inclusive educational environment where students with disabilities can thrive.

AI FOR STEM EDUCATION

A science class can now take students on a virtual journey through the solar system, allowing them to navigate planets and celestial bodies without leaving their desks. In a math lesson, analyzing large datasets becomes an interactive game rather than a daunting task. AI is transforming STEM education by making these scenarios not only possible but engaging, immersive, and highly personalized. Through AI-driven simulations, virtual labs, and intelligent tutoring systems, students can explore complex scientific concepts in ways that enhance both understanding and curiosity.

One of the most impactful applications of AI in STEM is its ability to simulate and visualize abstract scientific principles. Instead of simply reading about chemical reactions, students can interact with virtual chemistry labs, where AI adapts the experiments to their individual learning pace. In physics, AI-driven models can dynamically illustrate forces, motion, and energy transfer, allowing students to manipulate variables and observe real-time outcomes. These tools bring science to life, making it more accessible and less intimidating for learners of all skill levels.

AI is also revolutionizing the laboratory experience for schools with limited resources. Many institutions lack the funding or equipment for fully equipped science labs, but AI-powered virtual lab platforms now provide students with realistic, interactive experiment experiences. For example, AI-driven lab simulations allow students to conduct biology, chemistry, or physics experiments safely in a virtual environment, mimicking real-world lab conditions with remarkable accuracy. This not only eliminates accessibility barriers but also enables students to experiment without fear of failure, fostering a growth mindset and deeper inquiry.

AI plays a crucial role in engineering education by integrating robotics and coding into the curriculum. Students work with AI-powered robots, learning not only engineering principles but also machine learning and automation. This hands-on experience fosters a deep connection with the subject as students see the immediate impact of their coding and design efforts. Robotics projects also promote team collaboration, requiring students to design, build, and program robots that solve real-world problems. These activities build essential skills such as critical thinking, problem-solving, and teamwork, all of which are fundamental to success in STEM careers.

In mathematics, AI-powered data analysis tools help students visualize and interpret complex datasets, transforming raw numbers into interactive visual stories. Platforms like **Wolfram Alpha** or AI-driven graphing tools allow students to explore mathematical concepts dynamically, reinforcing learning through immediate feedback. AI-driven tutoring systems can also identify knowledge gaps in real time, adapting lessons to ensure personalized learning experiences that cater to each student's pace and understanding level.

Encouraging hands-on, project-based learning is another way AI is enriching STEM education. Platforms like **TinkerCAD** (for 3D design) and **Code.org** (for programming) provide students with interactive environments where they can create, innovate, and experiment with AI-powered guidance. These projects often culminate in presentations or demonstrations, reinforcing not just technical skills but also communication, creativity, and problem-solving abilities—essential traits for future STEM professionals.

Looking ahead, AI will continue to open new frontiers in STEM education. In bioinformatics, AI can help students analyze genetic data, offering insights into complex biological processes such as DNA sequencing and disease prediction. In environmental science, AI-powered simulations can model ecosystems and predict climate change impacts, giving students a global perspective on environmental challenges. In physics, AI-driven predictive modeling can introduce students to emerging technologies, from quantum computing to space exploration, sparking interest in cutting-edge research.

By continuously integrating AI into STEM education, educators can equip students not just with theoretical knowledge but with the practical, problem-solving, and analytical skills needed for the future. As AI continues to redefine industries, preparing students for AI-enhanced STEM careers will be critical to ensuring they remain competitive in an increasingly technology-driven world.

AI IN STEM EDUCATION CASE STUDY: THOMAS JEFFERSON HIGH SCHOOL FOR SCIENCE AND TECHNOLOGY, ALEXANDRIA, VIRGINIA

Thomas Jefferson High School for Science and Technology (TJHSST), a renowned institution specializing in science and technology education, has been at the forefront of integrating artificial intelligence (AI) into its STEM curriculum. By leveraging AI tools and fostering a culture of innovation, TJHSST aims to enhance student learning experiences and maintain its leadership in STEM education.

TJHSST offers a comprehensive curriculum that emphasizes STEM subjects, incorporating AI concepts into various courses. Students engage in projects that involve machine learning, data analysis, and robotics, providing hands-on experience with AI technologies.

Students who completed AP Computer Science can take additional courses to develop their own AI programs to tackle complex problems, such as creating algorithms to play games like Othello and solving multiple Sudoku puzzles rapidly. Additionally, the school has undertaken projects like TJREVERB, a 2U CubeSat equipped with AI capabilities, which was launched aboard SpaceX's CRS-26 mission and deployed from the International Space Station. The emphasis is on students developing custom AI solutions to address real-world challenges. Engaging students in practical AI projects, such as satellite development, enhances their understanding and interest in STEM fields. These experiences prepare students for future careers in technology and engineering by providing real-world problem-solving opportunities.

A major factor contributing to the school's STEM success is its Computer Systems Lab which houses advanced computing facilities, including a supercomputer, enabling students to work on complex AI algorithms and data-intensive projects. This access to high-performance computing resources has been instrumental in facilitating sophisticated AI research and experimentation. Additionally, the school's faculty comprises experts in various STEM fields, providing mentorship and guidance in AI projects. Collaborations with industry partners and higher education institutions further enrich the learning environment, offering students exposure to real-world AI applications and challenges.

Implementing AI projects, such as CubeSat development, required significant funding and technical resources. TJHSST addressed these challenges by securing partnerships with organizations like Orbital Sciences Corporation, which donated essential components and facilitated satellite launches. Collaborations with industry and academic partners are crucial for providing resources and expertise that support advanced AI projects. These partnerships enable schools to offer cutting-edge educational experiences that align with current technological advancements.

Through its commitment to integrating AI into STEM education, Thomas Jefferson High School for Science and Technology exemplifies how specialized institutions can lead in adopting innovative technologies to enrich student learning and maintain excellence in science and technology education.

LANGUAGE LEARNING AND AI

A classroom where students engage in real-time conversations with AI tutors, practicing languages with the fluency and precision of a native speaker, is now a reality. AI is transforming language acquisition by providing personalized learning experiences that adapt to each student's unique needs and learning style. AI-powered language tutoring apps like **Duolingo** tailor lessons based on a learner's progress, ensuring that new exercises build on existing knowledge. These platforms incorporate gamification techniques to keep learners engaged, turning language practice into an enjoyable, interactive experience rather than a monotonous task.

Speech recognition technology is another breakthrough in AI-driven language learning, particularly in pronunciation practice. Tools like **Rosetta Stone** and **ELSA Speak** leverage AI to analyze a learner's speech in real time, providing instant feedback on pronunciation, accents, and intonation. This immediate correction mechanism allows students to refine their speaking skills with precision, helping them build confidence in their ability to communicate in a new language. Unlike traditional methods, which often rely on instructor feedback that may be delayed or limited, AI-driven speech analysis ensures continuous, self-paced improvement.

The impact of AI in classroom language instruction is already being seen in places like Tokyo, where a school integrated AI-driven language tools into its curriculum. Students used immersive AI-powered simulations to practice conversations in real-world scenarios, from ordering food at a café to negotiating business deals. These AI-driven environments provided a safe and pressure-free space where students could practice speaking without the fear of making mistakes in front of their peers. In multilingual classrooms, AI-powered real-time translation tools like **Google Translate** and **Microsoft Translator** help students access content in their native language while gradually building proficiency in a new one. By breaking down language barriers, AI ensures that all students, regardless of their linguistic background, can participate fully in the learning experience.

AI also plays a crucial role in facilitating cultural exchange. Platforms like **Tandem** and **HelloTalk** connect language learners with native speakers worldwide, fostering an environment of mutual learning and cross-cultural understanding. These AI-powered language exchanges go beyond vocabulary and grammar —they provide insights into different cultures, idioms, and social norms, making language learning more authentic and immersive. This approach improves language skills and promotes empathy and global awareness, helping students develop a deeper appreciation for different perspectives.

Despite its advantages, AI-driven language learning is not without challenges. One of the biggest concerns is the accuracy of AI translation tools. While AI can process vast amounts of linguistic data, mistranslations still occur, leading to potential confusion or miscommunication. Developers continuously refine AI algorithms to improve contextual understanding, but human oversight remains essential. Additionally, cultural relevance is another challenge—language is deeply intertwined with history, traditions, and social norms, and AI alone cannot fully convey these nuances. Educators must supplement AI-driven lessons with real-world cultural context, ensuring students learn not just words and grammar but also the values and traditions that shape a language.

By blending AI-driven tools with human insight and cultural awareness, language learning can become more accessible, engaging, and effective. AI is not just revolutionizing how people acquire new languages—it's fostering deeper connections, breaking down barriers, and creating a more globally interconnected world.

AI IN LANGUAGE LEARNING CASE STUDY: PARKVIEW MIDDLE SCHOOL, ANKENY, IOWA

Parkview Middle School has embraced artificial intelligence (AI) tools to support English for Speakers of Other Languages (ESOL) students. By integrating AI technologies, the school aims to bridge language gaps and provide personalized learning experiences, enhancing educational outcomes for non-English-speaking students.

Their ESOL teacher employs **Magic School AI** to overcome language barriers, particularly for less common languages not covered by traditional translation services. This tool facilitates communication and understanding between students and educators, ensuring that language differences do not hinder learning. Their reading interventionist utilizes **ChatGPT** to create customized reading materials tailored to the specific needs and skill levels of her students. This personalization ensures that reading interventions are both relevant and effective, catering to individual learning requirements.

The use of AI tools enables the creation of individualized educational content, allowing students to engage with materials that are directly aligned with their learning needs and language proficiency levels. And, by addressing language barriers, AI facilitates better interaction between students and teachers, fostering a more inclusive and supportive learning environment.

Integrating AI into the classroom requires investment in technology and training. The school addressed this by prioritizing professional development and allocating resources to ensure effective implementation. Ongoing training for teachers is crucial to stay updated on emerging AI technologies and to integrate them effectively into pedagogical practices.

Interestingly, some educators were hesitant to adopt AI tools, concerned about potential misuse or overreliance by students. To mitigate this, teachers ensured that AI tools were used as supplements to traditional learning methods, not replacements, and provided guidance on responsible usage. Educators recognized the importance of guiding students on the ethical and effective use of AI tools, ensuring that technology enhances learning without compromising academic integrity.

Parkview Middle School's experience demonstrates that thoughtful integration of AI can significantly enhance language learning for non-English-speaking students, providing personalized educational experiences and supporting teachers in their instructional roles.

AI IN HOMESCHOOLING

Homeschooling has become an increasingly popular educational option, allowing families to tailor learning experiences to their children's unique needs. However, this approach has challenges, including curriculum design, access to specialized knowledge, and maintaining consistent engagement. AI is transforming the homeschooling landscape by addressing these challenges and providing innovative tools to enhance the learning process.

One of the most significant ways AI benefits homeschooling is through personalized learning paths. AI-powered platforms like **Khan Academy** and **Prodigy** adapt lessons based on a student's progress and performance. By analyzing how a student interacts with the material, these tools identify strengths and areas for improvement, tailoring content to meet individual needs. This ensures students learn at their own pace, fostering confidence and mastery in various subjects. For instance, a student struggling with fractions might receive additional exercises and explanatory videos, while a more advanced learner could be introduced to higher-level concepts. This adaptive approach eliminates the "one-size-fits-all" problem and creates a more effective and engaging educational experience.

AI-driven tools also help bring a new level of interactivity to homeschooling, making lessons more engaging and effective. Platforms like **Adventure Academy** and **Lingvist** provide multimedia-rich, gamified experiences that capture students' attention while teaching key concepts. Whether mastering a new language with **Lingvist** or exploring STEM concepts through immersive simulations, these tools keep students motivated and eager to learn.

For younger learners, gamified platforms like **Prodigy** turn math lessons into adventures. At the same time, older students can use tools like **Tinkercad** or **Blender** to create 3D models and animations, blending creativity with technology. This hands-on approach caters to diverse learning styles and helps students develop practical skills alongside theoretical knowledge.

AI can also simplify curriculum planning and resource organization, two critical aspects of homeschooling. Platforms like **Time4Learning** and **Homeschool Panda** provide parents with tools to create structured lesson plans, track progress, and access a wide range of educational resources. These systems offer analytics and recommendations, helping parents make informed decisions about their child's learning path. For example, **Homeschool Panda** includes features like collaborative planning and progress tracking, making it easier for parents to align their curriculum with educational goals. Platforms like **Edmentum** offer adaptive testing and analytics, ensuring students meet educational benchmarks while identifying areas that require additional focus. Other tools like **Homeschool Boss** automate scheduling, grading, and progress tracking, allowing parents to manage homeschooling more efficiently.

Finally, one common concern with homeschooling is limited peer interaction. AI-powered platforms like **Outschool** offer virtual classrooms and collaborative projects where students can connect and work with peers worldwide. These opportunities foster social skills and teamwork, addressing a key gap in traditional homeschooling.

CHAPTER 5: ETHICAL CONSIDERATIONS IN AI USAGE

A I in the classroom can seamlessly enhance learning and personalize education for every student while remaining an unobtrusive yet powerful force. Beneath this innovation, however, lies a complex web of ethical challenges that demand careful consideration. Educators stand at the intersection of technological progress and ethical responsibility, where the promise of AI must be balanced with the principles that protect the integrity of education.

Integrating AI into the classroom isn't just about adopting new technology—it's about ensuring that these tools align with core ethical values that preserve the humanity of learning. Autonomy, transparency, and accountability are the foundations of ethical AI implementation.

- **Autonomy** ensures that AI is a supportive tool rather than a decision-maker, empowering students and educators rather than dictating their paths.

- **Transparency** demands clarity and openness in how AI tools function, make recommendations, and influence learning outcomes.
- **Accountability** requires that educators and developers remain responsible for AI's impact on students, ensuring that its use enhances learning without compromising fairness, privacy, or integrity.

By navigating these ethical considerations, educators can harness AI's potential while safeguarding the values that define quality education—equity, trust, and student empowerment.

UNDERSTANDING AI ETHICS: KEY ISSUES FOR EDUCATORS

AI in education involves balancing innovation with responsibility. While AI offers unprecedented opportunities to enhance learning, it also poses ethical dilemmas that educators must address. Who should hold the decision-making authority—AI or educators? This question is central to discussions about AI's role in education. Educators need to maintain control, using AI as a tool to inform decisions rather than allowing it to dictate outcomes. Potential unintended consequences, such as reliance on AI-generated content without critical evaluation, underscore the need for ethical scrutiny. Consider a scenario where AI suggests educational pathways that unintentionally reinforce stereotypes or biases. Without human oversight, such consequences can slip through unnoticed, impacting student development.

The importance of ethical literacy in AI cannot be overstated. Educators must be prepared to face AI's ethical challenges—from bias and transparency to privacy and accountability. Ethical literacy involves understanding the principles that govern AI use and integrating these discussions into your curriculum. By facilitating classroom debates on AI ethics, you empower students to analyze, critique, and engage with AI thoughtfully. Encourage them to question AI's impact on their learning, privacy, and society at large, fostering a mindset that values responsibility, fairness, and ethical innovation. These discussions not only enhance critical thinking but also prepare students to interact with AI responsibly, ensuring that future generations recognize the ethical considerations embedded in technology use.

DATA PRIVACY: SAFEGUARDING STUDENT INFORMATION

A stark reality of today's world puts every student's personal information at risk of exposure simply because of the technology used in their education. This is why safeguarding student data in AI applications is critical. With AI, the stakes are high. Data breaches can have severe consequences, ranging from identity theft to unauthorized access to sensitive educational records. Understanding the importance of data privacy is not just about protecting information; it's about maintaining trust in the educational system. When student data is mishandled, the repercussions extend beyond the immediate breach, affecting reputations and relationships within the school community.

In educational settings, data privacy concerns are many. Collecting and storing personal information such as grades, health records, and behavioral data raises questions about who has access and how this data is used. Third-party access to student data is another pressing issue. Many AI tools are developed by external companies, making it essential to scrutinize their data-handling practices. Concerns arise when AI systems store conversations or interactions for training purposes, potentially sharing this information without proper consent. Such practices can inadvertently violate privacy regulations, putting schools at risk of legal repercussions.

Several strategies can be adopted to enhance data privacy. Encryption is a fundamental tool that ensures that data remains secure during transmission and storage. Secure data storage solutions, such as cloud-based systems with robust security measures, can protect information from unauthorized access. Regular audits of AI tools and their data usage practices are crucial. These audits help identify potential vulnerabilities and ensure compliance with privacy policies. Schools can also implement data access controls, restricting who can view or modify student data. By taking these steps, educational institutions can create a safer digital environment for students.

Legal and regulatory frameworks play a vital role in safeguarding student data. In the United States, the Family Educational Rights and Privacy Act (FERPA) governs the privacy of student records, ensuring that data is accessed and shared responsibly. FERPA mandates that schools obtain written consent from parents or eligible students before disclosing personal information. Similarly, the General Data Protection Regulation (GDPR) in the European Union requires organizations to obtain permission before collecting personal data and mandates stringent security measures for data protection. Compliance with these regulations protects student information and demonstrates a commitment to ethical data-handling practices. By adhering to these guidelines, schools can ensure they meet legal obligations while fostering a culture of trust and responsibility in handling student data.

IDENTIFYING AND MITIGATING AI BIAS

AI can be a helpful assistant in your classroom, analyzing student data and suggesting ways to enhance learning. It sounds promising, but what happens when this assistant only understands part of the story? That's where AI bias comes in. AI bias occurs when algorithms produce prejudiced results due to flawed data or design. This bias can skew educational outcomes, leading to unfair advantages or disadvantages for certain student groups. It manifests in various ways, such as recommending resources that favor one demographic over another or grading systems that fail to account for diverse learning styles. Understanding these biases is crucial because they can perpetuate existing inequities rather than leveling the educational playing field.

Bias in AI systems often originates from the data used to train them. If historical data reflects societal prejudices, AI may inadvertently perpetuate these issues. For example, if an AI tool is trained on data that predominantly features students from one cultural background, it might make recommendations that don't resonate with students from other backgrounds. Algorithmic design choices also play a role. Sometimes, these choices unintentionally prioritize specific outcomes over others, leading to biased results. Additionally, the decision-making processes embedded within algorithms can introduce biases if they rely on incomplete or unrepresentative data sets. These biases highlight the importance of being critical about the inputs that shape AI's outputs.

Mitigating AI bias requires deliberate and informed action. Regularly updating and diversifying data sets is a fundamental step. You can mitigate the risk of perpetuating biases by ensuring that AI systems are trained on comprehensive and inclusive data. Implementing bias detection tools is another effective strategy.

These tools can identify potential biases within AI systems, allowing for timely adjustments.

Additionally, transparency in algorithmic design is vital. Developers should document the decision-making processes within AI tools, enabling educators to understand and address any biases that might arise. This transparency fosters trust and ensures that AI applications align with educational goals.

As educators, you play a pivotal role in recognizing and addressing AI bias. Critical evaluation of AI tool recommendations is essential. Don't accept AI outputs at face value; question how these recommendations were generated and whether they align with your students' needs. When selecting AI tools, consider the diversity of data they are based on and the potential biases they might harbor. Engage with developers to ensure these tools are subjected to rigorous testing for bias. By taking these proactive steps, you create a fairer and more inclusive educational environment. Educators serve as the final checkpoint, where AI's potential meets the reality of diverse classrooms.

FRAMEWORKS FOR ETHICAL AI IMPLEMENTATION IN SCHOOLS

Bringing AI into the classroom isn't just about leveraging technology—it's about doing it responsibly. To ensure AI enhances education without compromising ethics, schools can use established frameworks that act like a moral compass, guiding decisions and practices.

One such resource is the **IEEE's Ethically Aligned Design**, which lays out principles focused on the well-being of both students and educators. It emphasizes key values like transparency and accountability, ensuring that AI systems are not just effective but also fair and understandable. Similarly, **UNESCO's AI Ethics Guidelines** champion the protection of human rights and dignity, promoting fairness and the need for human oversight. These aren't just lofty ideals—they're practical tools designed to help educators apply AI thoughtfully and responsibly in their schools.

Turn frameworks into action. A strategic approach is essential to making these ethical principles part of everyday school life. The first step? Developing school-wide AI ethics policies. These policies should reflect the core values from frameworks like IEEE and UNESCO, offering clear, actionable guidelines for how AI tools are used, monitored, and evaluated.

But policies alone aren't enough—training is key. Teachers, administrators, and even support staff need ongoing education about the ethical implications of AI. This isn't a one-and-done session; as AI technology evolves, so should the training, ensuring everyone stays informed about new developments and ethical considerations. By weaving ethical AI practices into the school's culture, educators create an environment where technology serves the students, not the other way around. This approach ensures that AI enhances learning while safeguarding student well-being.

Involve the whole school community. Ethical AI implementation isn't a solo mission—it's a team effort. Collaboration among teachers, administrators, students, and parents is crucial for developing meaningful AI ethics policies. When everyone's voice is heard, the resulting guidelines are more comprehensive and relevant to the entire school community.

Engaging with AI developers is also a critical piece of the puzzle. Schools should work closely with the creators of AI tools to ensure they meet ethical standards from the start. This partnership helps flag potential ethical issues early, preventing problems before they impact students and educators.

Keep it moving with continuous evaluation. The journey doesn't stop once AI tools are in place. Regular reviews of how AI affects student outcomes are essential to ensure these tools align with ethical guidelines and educational goals. By consistently assessing AI's impact, educators can spot areas for improvement and make adjustments as needed. This culture of continuous evaluation helps schools stay adaptable in the face of evolving AI technologies and ethical challenges.

Setting up feedback loops with stakeholders provides valuable insights, allowing schools to fine-tune their AI practices and policies. In the end, it's all about ensuring AI remains a positive, ethical force in education, supporting both teaching and learning in meaningful ways.

IMPROVING EQUITABLE ACCESSIBILITY TO AI TOOLS

Picture a classroom where every student, no matter their background or resources, has the chance to engage with and benefit from AI technology. This is a necessary goal that highlights the importance of equitable access to AI tools in education. However, the digital divide remains a major hurdle. This gap between students who have easy access to technology and those who don't can prevent learners in underresourced schools from tapping into the full potential of AI. Financial constraints, like the inability to afford modern devices and infrastructure issues, such as unreliable internet connections, create barriers that can leave many students behind in our increasingly digital world.

Addressing these challenges takes creative solutions and a steadfast commitment to equity. One practical approach is for schools to actively pursue grants and funding opportunities aimed at enhancing educational technology. These resources can help schools secure the necessary hardware and software to integrate AI tools effectively into classrooms.

Another powerful strategy is utilizing open-access AI platforms, which provide free or low-cost educational tools. These resources are particularly valuable for schools with tight budgets, offering a practical starting point to bring AI into the learning environment without overwhelming costs. Additionally, partnerships with tech companies can open doors to resources and training, helping bridge the technology gap and ensuring students gain the necessary skills.

SUCCESS STORIES FROM THE FIELD

Across the country, schools are proving that equitable access to AI tools is achievable with the right approach. In Texas, for instance, the Technology Lending Grant Program offers grants to school districts and charter schools to provide students with the equipment necessary to access and use digital instructional materials at school and at home. This program aims to bridge the digital divide by supplying devices like laptops and tablets to students in need.

Meanwhile, rural schools in Maine faced a different challenge: poor internet connectivity. The ConnectKidsNow initiative focuses on connecting Maine students to high-speed internet, especially in rural areas. The program aims to build the necessary infrastructure by collaborating with local internet providers to secured high-speed internet access for all students. This critical improvement allowed them to effectively incorporate AI tools into their curriculum, giving students in remote areas the same opportunities as those in more urban settings.

What stands out in these success stories is a common thread of strong leadership and community involvement. They show that self-determination, collaboration, and foresight can overcome barriers to ethical AI use. By prioritizing ethical considerations and involving all stakeholders, schools can leverage AI to enhance learning in meaningful ways.

CHAPTER 6: BUILDING A CULTURE OF ETHICAL AI USE

Each student's learning path must be guided by principles that put ethics at the forefront—where technology supports education without compromising core values. As AI becomes a more significant part of teaching practices, building a culture that prioritizes ethical use is more important than ever. This means setting clear standards and creating an environment where responsible AI use isn't the exception—it's the rule. Schools need to lead this transformation, ensuring that AI tools not only enhance learning but do so in ways that respect privacy, fairness, inclusivity, and personal integrity. By fostering this culture, educators can ensure technology remains a tool for empowerment, not exploitation.

CREATING A CULTURE OF ETHICAL AI USE IN SCHOOLS

The first step in building this culture is developing ethical standards for AI implementation. Schools should craft comprehensive guidelines outlining acceptable AI use, ensuring these align with educational goals and ethical principles. Think of this as a roadmap—a code of conduct that addresses crucial issues like data privacy, transparency, and accountability. Educators play a vital role here, monitoring AI tools and ensuring they meet these standards.

Ethical awareness is the cornerstone of responsible AI use. It's not just about rules—it's about understanding why they matter. Schools can raise this awareness among teachers, students, administrators, and parents. Hosting workshops on ethical AI for teachers and staff can demystify complex concepts, using real-world scenarios to highlight potential challenges. Imagine a teacher workshop where participants debate the ethical implications of using AI to grade essays—this kind of hands-on engagement makes ethics tangible and relevant. Similarly, information sessions for parents and students help build trust and understanding within the broader community.

Leadership is key to championing ethical AI use. School leaders set the tone, demonstrating their commitment through actions and decisions. Leadership training on AI ethics equips them with the tools to guide responsible AI implementation. These programs should cover everything from ethical decision-making processes to practical tips on managing AI tools. Real-life case studies of schools that have successfully navigated ethical AI challenges can offer valuable insights, highlighting both triumphs and pitfalls. By leading by example, school leaders can inspire their communities to prioritize ethical AI use.

Integrating ethics into the curriculum is another powerful way to build this culture. This goes beyond theoretical discussions—it's about embedding ethical considerations into everyday learning.

Technology classes, for example, can become platforms for exploring AI ethics, allowing students to engage with real-world dilemmas. Encouraging student-led discussions on AI ethics fosters a sense of agency and responsibility. Picture students debating the fairness of AI-driven grading systems or role-playing scenarios where they must navigate ethical decisions involving their use of AI tools.

Ethics modules can further deepen students' understanding, covering core concepts like transparency, accountability, and fairness. By making ethics a regular part of learning, schools ensure that students develop the ethical literacy needed to function in an AI-driven world. This approach prepares students for future careers and instills a lifelong commitment to ethical decision-making.

Building a culture of ethical AI use is a collective effort. It involves setting clear standards, raising awareness, fostering ethical leadership, and integrating ethics into the curriculum. By taking these steps, schools can create environments where AI enhances education while upholding the values that matter most.

ENGAGING PARENTS AND COMMUNITIES IN ETHICAL AI DISCUSSIONS

Parents and communities play a crucial role in shaping how AI is perceived and utilized in schools. Their support can significantly influence the success of AI initiatives. When parents understand the benefits and limitations of AI, they can become advocates for its use, helping to dispel myths and promote informed discussions. Engaging these stakeholders means inviting them into the conversation, making them partners in the educational process. This partnership requires openness and a willingness to listen to their concerns, ensuring that AI is used responsibly and ethically in the classroom.

To foster effective engagement, schools should consider hosting informational workshops and seminars specifically tailored for parents and community members. These events can demystify AI, explaining what it is and how it's being used to enhance education. Parents need clear, accessible resources to understand the impact of AI on their children's learning. Schools can provide a clearer picture of AI's role in education by offering hands-on demonstrations and interactive sessions. These events also allow parents to voice their concerns and ask questions, creating a dialogue that strengthens community bonds.

Addressing common concerns and questions is a key part of these discussions. Many parents worry about privacy issues and the ethical implications of AI in education. They may fear that AI could collect or misuse their children's data or replace important teacher-student interactions. Parents are also concerned about how students can responsibly use ChatGPT or other AI tools to do their homework. Schools must address these fears head-on, offering transparency about how AI tools are used and what safeguards are in place to protect students. Providing clear, concise explanations of data privacy policies and ethical guidelines can help alleviate these concerns, building confidence in the school's approach to technology.

Ongoing discussions about AI ethics can be facilitated through various tools and platforms. Online forums and community meetings enable continuous dialogue, allowing parents and educators to share insights and updates on AI initiatives. Schools might create dedicated sections on their websites or use social media to keep the community informed about AI developments. These platforms should be interactive, encouraging feedback and questions from parents. By keeping the lines of communication open, schools can ensure that parents feel involved and informed, contributing to a culture of transparency and trust.

Engaging parents and communities in these discussions has many benefits. Collaborative efforts can lead to more ethical AI practices as parents and educators work together to ensure that AI tools are used responsibly. This collaboration strengthens community-school relationships, fostering an environment where everyone feels invested in the educational process. As awareness and understanding of AI's impact grow, parents are more likely to support and advocate for its use, helping to create a positive, informed community that embraces technology as a tool for enhancing learning.

HELPING STUDENTS BECOME INFORMED, RESPONSIBLE USERS OF AI

In today's classrooms, AI tools like ChatGPT are becoming a staple for students working on assignments. While these technologies offer a wealth of resources, they also present challenges related to student honesty and accountability. Some of the more successful strategies include the following:

1. **Addressing AI Use in Homework Assignments**:

Statistics show that a growing number of students use AI to help with homework. This trend calls for a balanced approach in which AI is an aid, not a shortcut. Teachers can guide students in using these tools responsibly by setting clear expectations.

- **Clarify Expectations**: Educators should clearly define acceptable uses of AI tools. For instance, using AI for brainstorming or grammar checks may be allowed, while submitting AI-generated essays as original work is not. A word of caution: AI detection tools are reported to be notoriously unreliable and generally not recommended. Take care when using any AI detection system.
- **Promote Transparency**: Encourage students to disclose when and how they use AI in their assignments. This can be facilitated by including an "AI Usage Statement" section in submissions.
- **Develop Critical Thinking**: Require students to submit a reflection alongside AI-assisted work detailing how the tool was used and what insights were gained. This encourages students to engage critically with AI outputs rather than relying solely on technology.
- **Design AI-Resistant Assignments**: Create assignments that require personal reflection, local context, or specific classroom discussions that AI tools can't easily replicate.
- **Instill AI as an Aid, Not a Shortcut**: Teachers should emphasize that AI is a tool to assist learning, not replace critical thinking or personal effort. They should reinforce that mastering subjects requires engagement beyond what AI can provide.

2. **Developing Student Awareness Programs:**

Raising awareness about ethical AI use is crucial in fostering responsible behavior. Schools can implement various programs to engage students in understanding the implications of AI. Developing student awareness programs is essential for promoting a culture of responsible AI use.

- **AI Ethics Clubs:** Create AI ethics clubs or initiatives where students explore the ethical dimensions of AI in their daily lives. These programs can be student-led, empowering learners to take charge of their AI education and promote ethical usage among peers.
- **Student-Led Campaigns**: Encourage students to develop campaigns promoting responsible AI use. Campaigns might include creating posters highlighting responsible AI practices or organizing events where students discuss how AI impacts their learning. Such activities raise awareness and cultivate a sense of ownership and accountability.
- **Workshops and Seminars**: Host events with AI professionals to discuss ethical considerations and the responsible use of technology.

3. **Fostering Critical Thinking Skills:**

Critical thinking is essential in the AI era, equipping students to navigate complex information landscapes. Critical thinking involves analyzing information, questioning assumptions, and making decisions based on reason and evidence. In the context of AI, students must assess AI-generated content carefully, recognizing potential biases or inaccuracies. Students should be aware that AI tools can provide inaccurate, biased, or incomplete information.

- **Avoid Blind Acceptance**: Teach students to see AI as a starting point, not the final authority. Critical skills enable students to question AI outputs, verify information, and avoid unquestioningly accepting AI-generated responses.

- **Question AI Outputs**: Encourage skepticism and inquiry into AI responses. AI output can provide incorrect or misleading information, known as chatbot hallucinations. Students should be aware of this and question all AI responses.
- **Compare to Trusted Sources**: Have students validate AI information against reputable databases or academic resources.
- **Check for Bias**: Discuss how AI models can reflect biases in training data. Students can be tasked with identifying biased language in AI-generated news articles or evaluating the objectivity of AI-written essays. This exercise teaches them to approach AI tools critically, fostering a mindset that questions and investigates rather than accepts at face value.
- **Analyze Reliability**: Teach students to evaluate the credibility of AI-generated facts, especially in subjects like history or science.

4. **Incorporating Critical Thinking and Digital Literacy into the Curriculum:**

Critical thinking and digital literacy can be integrated across subjects. Students should practice comparing AI-generated essays with trusted sources to assess accuracy and reliability. By engaging in this evaluative process, they learn to discern the quality of AI outputs and develop a deeper understanding of the subject matter.

- **History:** Analyze primary sources and compare AI-generated historical summaries for accuracy.

- **Science:** Evaluate AI-assisted data analysis in experiments, questioning anomalies or inconsistencies.
- **Interdisciplinary Projects**: Design projects that use AI and require students to critically assess its output are particularly effective. For instance, a project might involve students analyzing AI-generated summaries of scientific articles, comparing them to human-written versions, and presenting their findings. Such projects enhance critical thinking and demonstrate the practical application of these skills in real-world contexts.

5. **Recognizing and Questioning Manipulated Media:**

AI-powered tools can create deepfakes and edited images, making media literacy vital. Students must also develop the critical skill of recognizing and questioning manipulated media, such as deepfakes or edited images.

- **Media Literacy Education**: Teach students to identify signs of manipulated media and verify sources. AI-powered manipulation can blur the line between reality and fiction, making it difficult to distinguish authentic content. To counter this, media literacy must be a priority in digital education. Students should learn techniques to identify altered media, such as checking for inconsistencies or using verification tools. Discussing the ethical implications of manipulated media can further enhance their understanding. By equipping students with the skills to navigate these challenges, educators help them become savvy digital citizens capable of discerning truth from manipulation.

- **Critical Analysis Exercises**: Use real-life examples of deepfakes to practice discerning authentic from altered content.

6. **Understanding AI-Driven Marketing and Content Recommendations:**

AI shapes online experiences through personalized marketing and content algorithms. AI-driven marketing and content recommendations shape much of what students see online. Understanding how these algorithms work is essential for becoming critical consumers of digital content.

- **Algorithm Education**: Explain how social media platforms use AI to recommend content, influencing what students see. Educators can explain how social media algorithms prioritize specific posts or how recommendation systems suggest videos based on viewing history. By exploring these concepts, students gain insights into the hidden mechanisms influencing their online experiences. Discussing the impact of targeted advertising can also highlight the ethical considerations of data usage. This awareness encourages students to think critically about the content they consume, fostering a more informed and reflective approach to digital interaction.
- **Critical Consumption**: Encourage students to question the motives behind recommended content and recognize echo chambers.

In conclusion, helping students become informed, responsible users of AI requires a multifaceted approach. It involves instilling critical thinking skills, fostering digital literacy, and encouraging ethical awareness. By adopting these strategies, educators can prepare students to navigate the complexities of an AI-driven world with confidence and integrity. As we look ahead, the next chapter will explore how AI can enhance classroom efficiency, offering practical solutions to streamline teaching practices and improve educational outcomes.

CHAPTER 7: AI CHALLENGES AND OVERCOMING OBJECTIONS

Amid all the promise and excitement for AI, there's a current of apprehension. Some educators worry that AI might be the harbinger of change that sidesteps their roles, overshadowing the personalized touch teachers bring to the classroom. This fear isn't unfounded. The rise of AI in various sectors, from healthcare to transportation, has reshaped job landscapes, and education is no exception. As AI tools like ChatGPT and adaptive learning platforms become commonplace, it's normal to feel a twinge of skepticism about what this means for traditional teaching.

ADDRESSING SKEPTICISM AND FEAR OF AI

The root of this skepticism often lies in concerns about job displacement. Educators, who have always been the cornerstone of learning, may fear being replaced by a machine that can seem as if it knows everything. This worry stems from misunderstanding AI's role. AI is not here to replace teachers but to augment their capabilities. It can automate repetitive tasks like grading or attendance, allowing educators to focus on what truly matters—teaching and engaging with students. Yet, the fear of technological domination persists, fueled by misconceptions and the rapid pace of AI development. This anxiety can be compounded by past experiences with educational technologies that promised much but delivered little.

Another source of skepticism is the belief that AI could lead to losing the human element in teaching. Education thrives on connection, empathy, and understanding, qualities AI lacks. Teachers are more than just knowledge providers; they are mentors, motivators, and confidants. The worry is that AI might lead to a sterile learning environment where data-driven decisions overshadow the nuanced insights educators bring. These concerns are valid. AI's analytical prowess cannot replace the warmth and encouragement that teachers offer. This fear highlights the need for AI to be integrated thoughtfully, ensuring that it complements rather than replaces the personal touch of teaching.

The fear of AI overtaking traditional teaching methods is often rooted in a lack of understanding of how these technologies function. Many educators feel unprepared to navigate the complexities of AI, fearing that their lack of technical expertise might render them obsolete. This fear is exacerbated by the rapid pace of AI advancements, which can overwhelm even the most tech-savvy. The challenge lies in bridging this knowledge gap and empowering teachers with the skills and confidence to use AI effectively. Training programs and professional development opportunities can be crucial in demystifying AI, turning it from a source of fear into a valuable tool.

Concerns about AI's potential to propagate bias are also prevalent. AI models are trained on vast datasets, which can inadvertently reflect societal biases. This can lead to skewed outcomes, affecting student assessments and learning experiences. Educators worry about the ethical implications of relying on AI tools that might not be as impartial as they appear. Addressing this skepticism requires transparency about how AI systems function and the measures in place to mitigate bias. Schools can foster a sense of ownership and trust by involving educators in developing and evaluating AI tools, reducing fears about biased outcomes.

Finally, there's the concern that AI might lead to a reevaluation of educational goals. With AI handling informational tasks, the focus should shift from teaching students what to learn to how to learn. This reframing can be daunting, challenging longstanding educational paradigms. Educators may feel unsure how to adapt their teaching methods to fit this new landscape. However, this shift also presents an opportunity to foster critical thinking and problem-solving skills, preparing students for a future where AI is integrated into many aspects of life and work. By embracing this change, educators can equip students with the skills they need to navigate an AI-driven world confidently.

Understanding these sources of skepticism is the first step in overcoming the fear of AI in education. By addressing these concerns and fostering open dialogue, educators can transform apprehension into acceptance, embracing AI as a valuable ally in enhancing student learning.

THE CHALLENGE OF LOW-RESOURCE ENVIRONMENTS

In many schools, especially those in rural or economically disadvantaged areas, the prospect of integrating AI into education can feel like a distant dream. The barriers are real and formidable. Schools often lack funding, outdated infrastructure, and inadequate access to the necessary hardware to support AI tools. These challenges can create a sense of inequality, where the technological advancements available to some are out of reach for others. This disparity affects not only the quality of education students receive but also their readiness for a future where technology plays a crucial role. The digital divide is more than just a lack of devices; it's about the opportunities students miss when their schools can't afford the latest educational technology.

However, resourceful solutions exist for schools willing to think creatively. One practical approach is utilizing free or low-cost AI tools and platforms. There are numerous AI applications available at no cost that can still significantly enhance the educational experience. For instance, tools like **ChatGPT** can assist with lesson planning and generating writing prompts, offering educators a helping hand without financial strain. **Brisk**, a Chrome extension, adapts articles for different proficiency levels, making it easier for teachers to differentiate instruction. By leveraging these free resources, schools can begin to integrate AI into their classrooms without overextending their budgets.

Partnering with local tech companies for resource sharing is another innovative strategy. These partnerships provide schools access to AI tools and expertise that might otherwise be unavailable. Tech companies often have a vested interest in education, recognizing it as a means to cultivate future talent. By collaborating with these companies, schools can secure the tools and the training needed to use them effectively. This collaboration might involve workshops, mentorship programs, or even grants to support technology integration. Such partnerships can be mutually beneficial, as companies gain a deeper understanding of educational needs while schools receive valuable resources.

Several schools have successfully bridged the technology gap by employing community-driven initiatives. For example, a school district in Arizona launched a crowdfunding campaign to raise funds for AI projects. The community rallied behind the initiative, understanding the long-term benefits of investing in their children's education. With the funds raised, the district was able to implement AI-driven learning programs that personalized instruction and improved student engagement. This example highlights the power of community involvement in overcoming financial constraints and underscores the importance of transparent communication about the benefits of AI in education.

Accessing funding for AI initiatives requires strategic planning and persistence. Educators can explore various grant opportunities offered by government agencies, nonprofits, and private foundations focused on educational technology. It's essential to thoroughly research each grant, understanding its criteria and alignment with the school's goals. Writing a compelling grant application is crucial. Highlight the potential impact of AI on student learning and the innovative ways the school plans to implement these tools. Providing a clear, detailed plan with measurable outcomes can strengthen the application, increasing the likelihood of securing funding.

Navigating grant applications can be daunting, but resources are available to guide educators through the process. Many organizations offer workshops or online courses on grant writing, providing tips and techniques to craft persuasive proposals. It's also beneficial to connect with other educators who have successfully obtained grants. Sharing insights and experiences can offer valuable perspectives and encouragement. Additionally, leveraging school or district networks to support grant applications can provide a collaborative approach, pooling resources and expertise to enhance the chances of success.

In low-resource environments, integrating AI into education can be challenging but not impossible. By embracing innovative solutions and seeking out partnerships and funding opportunities, schools can begin to narrow the digital divide. The goal is to create an equitable educational experience where all students can access the technology tools they need to thrive. These efforts, while demanding, are a crucial investment in the future, ensuring that every student, regardless of their background, has the opportunity to succeed in an increasingly digital world.

THE ISSUE OF TIME MANAGEMENT: INTEGRATING AI WITHOUT OVERLOAD

Teaching is a demanding profession, and let's face it, there are never enough hours in the school day. Between lesson planning, grading, meetings, and countless other responsibilities, adding AI integration to your to-do list might feel like an insurmountable task. It's a common concern among educators. How can you fit AI into an already-packed schedule without feeling overwhelmed? The key lies in smart time management and gradual integration. Recognizing the constraints you face is the first step toward finding solutions that work. It's about making AI a part of your routine, not an additional burden.

One practical way to manage time effectively is to start by identifying AI tools that can save you time. Think of applications designed to handle routine tasks like attendance tracking or grading. AI-driven platforms like **Education Copilot** can help automate these processes, freeing you to focus on teaching. For instance, an AI attendance system can streamline roll calls, giving you more time to engage with students. Similarly, AI tools like **Gradescope** can significantly reduce grading time by automating the assessment of multiple-choice and even short-answer questions. By embracing these tools, you can reclaim hours each week, allowing you to invest more energy into lesson planning and student interaction.

But integrating AI isn't just about choosing the right tools; it's also about how you introduce them. A gradual implementation plan can ease the transition, ensuring you don't feel overwhelmed. Start small by incorporating AI into one aspect of your teaching, such as lesson planning or grading. As you become more comfortable, expand its use to other areas. This phased approach allows you to learn and adapt at your own pace, minimizing stress and maximizing effectiveness. Consider setting aside time each week to explore new AI features and assess their impact on your workload. It's about building familiarity and confidence, turning AI from a daunting prospect into a reliable ally.

Prioritization and planning are crucial when integrating AI into your teaching practices. Begin by setting clear goals for what you want to achieve with AI. Are you looking to improve student engagement, streamline administrative tasks, or personalize learning experiences? By identifying your priorities, you can focus on AI initiatives that align with your objectives. Once you've set your goals, create a realistic timeline for adoption. Remember, it's not a race. Take the time to explore and experiment with AI tools, ensuring they meet your needs and enhance your teaching. Break tasks into manageable steps, and celebrate small victories along the way. This approach reduces the pressure and maintains your enthusiasm for innovation.

Incorporating AI into your classroom doesn't mean sacrificing the personal touch that makes your teaching unique. Instead, view AI as an extension of your capabilities. It can provide insights and support, allowing you to focus on building meaningful connections with your students. The integration process should be collaborative, involving students in the journey. Encourage them to engage with AI tools, offering feedback on their experiences. This provides valuable insights and fosters a sense of ownership and excitement among students. As they see the benefits of AI in their learning, they'll become more invested in the process, enhancing the overall educational experience.

To manage time effectively while integrating AI, consider setting aside dedicated periods for professional development. This could involve attending workshops, webinars, or collaborating with colleagues to share insights and strategies. Learning from others who have successfully integrated AI can provide inspiration and practical tips. It's about building a supportive network where you can exchange ideas and troubleshoot challenges together. By prioritizing professional development, you ensure you remain informed and equipped to make the most of AI in your classroom.

Finally, remember that integrating AI is not about perfection. It's about progress. Mistakes and challenges are part of the process, and each offers an opportunity to learn and grow. Embrace the journey with an open mind, and don't hesitate to seek support when needed. Whether it's reaching out to colleagues, joining online educator communities, or connecting with AI experts, a wealth of resources is available. By managing your time wisely and approaching AI integration with a strategic mindset, you can enhance your teaching practice without feeling overwhelmed.

OVERCOMING RESISTANCE: BUILDING A PRO-AI CULTURE

Excitement about AI doesn't happen overnight. It starts with cultivating a positive mindset that embraces change and innovation. We need to foster an environment where teachers feel encouraged to try new things. Let's celebrate the courage it takes to experiment with AI tools. Maybe you could introduce an "AI Innovator of the Month" award. Recognizing those who take the leap can inspire others to step out of their comfort zones. Celebrations like these can create a ripple effect, gradually shifting the school culture toward one that values innovation and creativity.

Creating channels for open dialogue is crucial. Regular meetings discussing AI initiatives provide a platform for educators to share their progress and experiences. This setup fosters a sense of community, where teachers feel supported in their efforts to integrate technology. It's important to provide a safe space where trial and error are accepted and encouraged. When educators know they can experiment without the fear of failure, they are more likely to explore the full potential of AI tools. These discussions can also highlight AI's real-world benefits, such as improved student engagement and personalized learning experiences.

Engaging all stakeholders—teachers, students, and parents—is another vital step. Hold workshops for parents to explain AI's role in education. When parents understand how AI enhances learning, they are more likely to support these initiatives. Students, too, should be involved in the conversation. Encouraging them to participate in AI projects or voice their opinions on AI tools can foster a sense of ownership and excitement. A collective buy-in from the entire school community can drive successful AI integration, as everyone works toward the common goal of enhancing education through technology.

Addressing resistance with empathy is key. There will always be concerns, and it's essential to understand and address them thoughtfully. One-on-one meetings can be effective in reassuring hesitant colleagues. These conversations allow you to listen to individual concerns and provide tailored solutions. Highlight AI's benefits, such as reducing workload through automation or offering insights into student progress. Focusing on these advantages can help alleviate fears and build confidence in AI's role as a supportive tool rather than a threat.

Learning from schools that have successfully adopted AI can provide valuable insights. Case studies offer concrete examples of overcoming obstacles and achieving success. For instance, a school that faced initial resistance might have succeeded by gradually introducing AI tools, aligning them with existing curricula, and continuously training staff. Strong leadership commitment can make a significant difference. Leaders who champion AI initiatives and provide necessary resources and support can inspire their teams to embrace technology. These stories serve as blueprints for other educators, offering practical lessons they can apply in their own contexts.

Analyzing key success factors from these case studies is crucial. Community involvement often stands out as a significant element. Schools that engage with their local communities, whether through partnerships or collaborative projects, fare better. Creative fundraising efforts can also support AI projects. Whether organizing community events or applying for grants, these initiatives can provide the financial backing needed to implement AI tools effectively. Overcoming resistance requires a mindset of perseverance. Sharing motivating stories of transformative change can inspire educators to push through challenges. It's about celebrating small victories and focusing on the long-term benefits AI can bring to education.

As we consider the steps needed to build a pro-AI culture, remember that change doesn't happen overnight. It requires a concerted effort from everyone involved. By cultivating a positive mindset, engaging stakeholders, and learning from successful examples, you can create an environment where AI is welcomed as a valuable ally in the educational landscape.

CHAPTER 8: BUILDING AI LITERACY AMONG EDUCATORS

Picture standing in front of a classroom where your students engage and actively participate in their learning. They use AI tools with ease, as naturally as they would a pen or a textbook. This scenario is within reach, but to make it a reality, we must first focus on building AI literacy among educators. In today's fast-paced educational environment, understanding AI is not just an advantage—it's essential. Providing educators with the right knowledge and tools can revolutionize teaching methods and elevate learning experiences for all. This chapter explores how we can set the foundation for this transformation.

AI TRAINING PROGRAMS FOR TEACHERS: GETTING STARTED

To start building AI literacy, establishing program foundations is crucial. Successful AI training programs must address the specific needs and objectives of educators. This begins with identifying what teachers need most—be it understanding basic AI concepts or applying AI tools in their classrooms. The curriculum should focus on practical applications, ensuring educators can see the immediate benefits of integrating AI into their teaching. Courses like *"AI Literacy for Educators"* from Teachers College Columbia University offer a structured approach, covering foundational AI skills and ethical awareness, making them an excellent model from which to draw inspiration.

Implementing training modules is the next step in this journey. These modules should be structured to gradually build AI literacy, beginning with the basics and progressing to more intermediate applications. Online courses and interactive webinars provide flexibility, enabling teachers to learn at their own pace. Resources like the free *"AI 101 for Teachers"* series by **Code.org** and partners are invaluable, offering foundational sessions that demystify AI and explore its responsible implementation. By providing multiple formats, educators can choose the learning method that best suits their needs, ensuring a more personalized and effective educational experience.

Evaluating training effectiveness is essential to ensure that AI literacy programs meet their objectives. Pre- and post-training assessments can gauge the knowledge gained, highlighting areas of strength and those needing further development. Feedback surveys offer insights into participant satisfaction, helping refine future training sessions. These evaluations are not just about measuring success; they're about continuous improvement. Educators can adapt and enhance their AI literacy efforts by understanding what works and what doesn't, ensuring they remain relevant and impactful. This iterative approach fosters a culture of learning and adaptation, which is essential for navigating the ever-evolving landscape of AI in education.

Building a strong foundation of AI literacy among educators is more important than ever. By establishing comprehensive training programs, implementing structured learning modules, and evaluating their effectiveness, we can empower educators to embrace AI confidently and creatively. This commitment to learning and growth will not only enhance teaching practices but also enrich students' educational experiences, preparing them for a future where AI is an integral part of their lives.

PROFESSIONAL LEARNING COMMUNITIES AND AI: SHARING KNOWLEDGE

Imagine a group of educators gathered in a classroom, passionately discussing the latest AI tools and how they can be leveraged to enrich student learning. This is the heart of a Professional Learning Community (PLC) focused on AI. These groups offer a dynamic environment where teachers can explore AI concepts together, share experiences, and develop strategies that work in real classrooms. By forming AI-focused educator groups within schools, you create a space where innovation thrives. Whether sharing the latest AI app or discussing challenges and triumphs, these communities foster a culture of collaboration that enhances AI literacy.

To elevate the impact of PLCs, it's essential to encourage collaborative learning. This means promoting peer-led study groups where educators can dive deep into AI applications, supported by colleagues who share their passion. Collaboration doesn't stop at the school gates. Engaging with regional or national AI educator networks broadens perspectives and introduces a wealth of collective knowledge. Such networks can host workshops or webinars, offering diverse viewpoints and teaching methods. By participating in these communities, educators gain access to a broader range of experiences and resources, enriching their own teaching practices and expanding their understanding of AI in education.

Digital platforms are crucial in facilitating knowledge sharing within these communities. Platforms like **Google Classroom** allow educators to share resources, initiate discussions, and collaborate on AI projects, breaking down geographical barriers. Online forums and discussion boards allow teachers to ask questions, share insights, and seek advice from peers. These platforms offer a shared knowledge repository where all members can store and access lesson plans and case studies. By leveraging these digital tools, educators can maintain an ongoing dialogue about AI, ensuring that learning and collaboration continue beyond physical meetings.

Assessing the effectiveness of PLCs in enhancing AI literacy is crucial for understanding their impact. Tracking participation and engagement levels provides insights into how invested educators are in these communities. Surveys and feedback sessions can highlight areas of improvement, ensuring that PLCs remain relevant and responsive to educators' needs. Providing case studies of successful PLC initiatives can serve as inspiration, demonstrating how collaborative efforts have led to tangible improvements in AI literacy and classroom integration. These stories motivate current members and attract new participants, further strengthening the community.

Professional learning communities offer a powerful way to enhance AI literacy among educators. By forming these groups, promoting collaborative learning, utilizing digital platforms, and measuring their impact, you create an environment where educators can explore AI's potential together. This collaborative approach enriches individual teaching practices and fosters a collective commitment to innovation and excellence in education. Through these communities, educators can confidently navigate AI's complexities, transforming their classrooms into dynamic spaces where technology and learning intersect.

HANDS-ON WORKSHOPS: LEARNING AI THROUGH PRACTICE

Hands-on workshops offer a unique opportunity to dive into the world of AI, providing practical experiences that go beyond theory. By designing workshops that focus on interactive AI learning, educators can truly grasp the power of these tools. Demonstrations of AI applications, such as real-time data analysis or interactive simulations, can illuminate how these technologies can be integrated into everyday teaching. Participants can see firsthand how AI transforms abstract concepts into tangible learning experiences, making lessons more engaging and impactful for students.

Experiential learning is the heart of these workshops, emphasizing the importance of learning by doing. Imagine educators working on AI projects that can be directly applied in their classrooms. These projects might include developing lesson plans using AI to tailor content to different learning styles or creating interactive activities that encourage student participation. Problem-solving activities using AI technologies can also be a highlight. For example, teachers could work on real-world scenarios where AI helps identify patterns in student performance, offering insights that inform teaching strategies. This hands-on approach builds confidence and inspires creativity, empowering educators to think outside the box and explore new teaching methodologies.

Access to the right tools and resources is crucial for the success of these workshops. Participants should be able to experiment with AI software, exploring free trial licenses that allow them to test functionalities without financial commitment. Providing AI hardware, such as tablets and sensors, can add another layer of interaction, helping educators visualize how these tools can be integrated into their teaching practice. This access ensures that teachers aren't just passive learners but active participants, fully immersed in exploring AI's potential. With the right resources, educators can leave the workshop with a toolkit of ideas and practical skills to implement immediately.

Assessing the outcomes of these workshops is key to understanding their impact. Post-workshop feedback and reflection sessions provide valuable insights into what participants found most beneficial and where they encountered challenges. This feedback helps refine future workshops, ensuring they remain relevant and effective. Conducting skill assessments can also highlight the competency gains achieved through the sessions. By evaluating these outcomes, educators can ensure that the workshops are not just one-off events but part of a broader strategy to enhance AI literacy. These evaluations foster a culture of continuous improvement, encouraging educators to keep exploring and refining their AI skills.

Through hands-on workshops, educators can unlock AI's potential in a supportive and engaging environment. By focusing on interactive learning experiences, providing access to essential tools, and assessing outcomes, these workshops empower teachers to embrace AI with confidence and creativity. This practical approach enhances individual teaching practices and contributes to a broader culture of innovation and excellence in education. With each workshop, educators move closer to a future where AI is seamlessly integrated into classrooms, enriching students' learning experiences and preparing them for the challenges of tomorrow.

LEVERAGING ONLINE RESOURCES FOR CONTINUOUS AI EDUCATION

In today's digital age, the wealth of online resources available for AI education is astonishing. As educators, tapping into these resources can significantly enhance your understanding and implementation of AI in the classroom. Start by curating a list of high-quality online education materials that cater specifically to AI. Massive Open Online Courses (MOOCs) such as **Coursera** and **edX** offer comprehensive courses on AI, ranging from beginner levels to more advanced topics. Educational AI blogs provide a steady stream of the latest developments and insights, while YouTube channels dedicated to AI can offer visual and practical demonstrations of AI tools in action. Facebook groups like **AI for Education** create communities where you can engage with other educators, share experiences, and ask questions in a supportive environment.

Self-directed learning is a powerful tool in the journey to mastering AI. By setting personal learning goals, you can tailor your educational experience to fit your needs and interests. Maybe you're curious about how AI can streamline classroom management, or you're interested in developing AI-driven lesson plans. Whatever your focus, online courses provide the flexibility to explore these areas at your own pace. Joining online study groups can also offer additional support and motivation. These groups allow you to connect with peers who are on a similar path, enabling you to share resources, discuss challenges, and celebrate successes together. The key is to take charge of your learning, using these resources to build a personalized AI education pathway.

The accessibility and flexibility of online learning make it an attractive option for busy educators. With resources available anytime, anywhere, you can fit learning into your schedule without the constraints of traditional classroom settings. Online courses often offer a range of materials suited to different skill levels, ensuring that you can find content that matches your current understanding and pushes you to grow. This adaptability allows you to progress at a comfortable pace, revisiting complex topics as needed and diving deeper into areas that spark your interest. The freedom to learn on your terms is a significant advantage, making online resources a valuable tool in your professional development arsenal.

Resource List: Top Online AI Education Resources

Explore the following resources to kickstart your AI learning journey:

- Coursera's "AI for Everyone" course provides a broad overview of AI concepts.
- edX offers specialized AI courses from top universities.
- "IBM AI Education" is a collection of free, on-demand webinars for K-12 educators.
- The "AI for Education" Facebook group connects you with a community of educators and access to their webinars.
- The "AI in Education" blog offers insights and updates on AI trends.

- YouTube channels like "AI Explained" provide visual tutorials and demonstrations.

By leveraging these resources, you can enhance your AI literacy, equipping yourself with the knowledge to transform your teaching practices and inspire your students.

FROM THEORY TO CLASSROOM PRACTICE

Translating AI theory into practical teaching applications is where the magic happens. It's one thing to understand the concepts of AI, but it's entirely different to see those concepts come to life in the classroom. As educators, integrating AI into lesson plans begins with understanding how these technologies can enhance learning. For instance, AI algorithms can analyze student performance data to tailor lessons that meet individual needs. This practical application of AI theory allows you to provide personalized learning experiences that can significantly improve student engagement and outcomes. By embedding AI into your teaching practice, you create a dynamic classroom environment where technology supports and enriches every student's learning journey.

Getting from theory to practice involves developing practical implementation plans. This means creating step-by-step guides that effectively outline how to incorporate AI into your teaching methods. Start by identifying specific areas in your curriculum where AI can make a meaningful impact. For example, AI can be used to automate grading, freeing up time for more personalized instruction. Set realistic timelines to implement these changes, ensuring that you allow sufficient time for experimentation and adjustment. Iterative improvement is key; assess the outcomes and refine your approach based on what works best for your students. This process builds your confidence in using AI and encourages a culture of continuous innovation in your teaching practice.

Showcasing success stories can be a powerful motivator. Hearing from educators who have successfully integrated AI into their classrooms provides real-world examples of what is possible. Take the case of a high school teacher who used AI to create a custom tutoring system for math students. By analyzing data from student assessments, the system provided targeted exercises and feedback, resulting in improved test scores and increased student confidence. Testimonials from educators who have embraced AI highlight the challenges and rewards of this journey, offering valuable insights and inspiration. These stories demonstrate that integrating AI into education with the right tools and mindset can lead to transformative outcomes.

Adopting a growth mindset is crucial when it comes to AI adoption. This means viewing challenges as opportunities to learn and grow rather than obstacles to be avoided. Overcoming initial resistance or fear of AI is part of this mindset shift. Embrace the idea that AI is a tool to enhance, not replace, your teaching. By staying curious and open to new experiences, you can discover innovative ways to use AI that align with your educational philosophy. This approach fosters resilience and adaptability, empowering you to navigate the ever-changing landscape of educational technology with confidence and creativity.

Networking and mentorship are invaluable resources for educators exploring AI. Connecting with experienced AI mentors can provide guidance and support as you integrate these technologies into your classroom. Look for networking events or conferences focused on AI in education, where you can meet like-minded educators and industry experts. These events offer opportunities to share ideas, learn from others, and build a community of support. Establishing mentorship programs within your school or district can also facilitate knowledge sharing and collaboration, helping educators at all levels grow their AI literacy and confidence.

Incorporating AI into classroom practice is a journey that begins with understanding theory but truly comes alive through practical application. Educators can effectively integrate AI into their teaching methods by developing actionable plans, learning from successful examples, adopting a growth mindset, and connecting with supportive networks. This approach enhances educational outcomes and prepares students for the future by equipping them with the skills they need to thrive in a technology-driven world.

CHAPTER 9: AI TOOLS AND RESOURCES

In a classroom empowered by AI, technology is both an assistant and an amplifier of your teaching skills. The secret to this transformation lies in building a robust AI toolkit tailored to your unique classroom needs. This chapter explores how you can create a set of indispensable AI tools that will elevate your teaching to new heights.

BUILDING YOUR AI TOOLKIT: ESSENTIAL RESOURCES

Building an AI toolkit starts with identifying the core tools that can significantly impact your teaching practice. A versatile toolkit is essential because no two classrooms are the same. You need diverse tools to address the varied teaching and learning needs that arise. This diversity ensures that you can adapt to different teaching styles and classroom dynamics, creating a personalized experience for each student.

Compiling a custom AI toolkit involves careful selection and evaluation. Start by assessing the compatibility of potential tools with your existing systems. Consider how each tool integrates into your current workflow and whether it complements your teaching style. Usability is key; look for intuitive and easy-to-navigate tools, minimizing the learning curve for both you and your students. Balancing functionality with simplicity is crucial. A tool that offers a wide range of features is valuable, but only if those features are accessible and enhance your teaching rather than complicating it. Gather input from colleagues who have used these tools, as peer recommendations can provide valuable insights into what works well in an actual classroom setting.

Explore online resource libraries and educational technology conferences to expand your AI toolkit. Platforms like "**Ditch That Textbook**" offer curated lists of free and paid AI tools, providing a wealth of options to consider. These resources help you stay informed about the latest developments in educational AI, ensuring your toolkit remains relevant and effective. Attending conferences and workshops is another excellent way to discover new tools and learn from educators who successfully integrate AI into their teaching practices. Networking with other educators and sharing experiences can inspire fresh ideas and approaches, enriching your toolkit and enhancing your teaching.

LEVERAGING FREE AND PAID AI TOOLS EFFECTIVELY

When it comes to AI tools in education, both free and paid options have their place, offering distinct advantages and limitations. Free tools provide a fantastic entry point for educators just beginning to explore AI. They allow you to experiment without financial commitment, which is crucial when budgets are tight. However, free versions often come with limitations. They may have restricted functionality or lack the robust support to make a real difference in a classroom setting. For instance, a free AI tool might offer basic analytics but charge for more detailed reports that could help you tailor instruction more effectively. The key is to weigh these limitations against your needs and expectations.

On the other hand, paid AI tools often bring enhanced features and dedicated support. They can offer more comprehensive analytics, customizable templates, and integration capabilities that free tools might not provide. The cost-benefit analysis of investing in these tools involves considering how their advanced features can enhance your teaching and whether they justify the expense. Scalability is another factor; if you anticipate expanding your use of AI, a paid tool might offer the flexibility and resources necessary for growth. The decision to go with a free or paid tool should consider your class size, specific needs, and long-term educational goals.

Strategies for maximizing free AI tools involve creativity and resourcefulness. You can often combine multiple free tools to create a comprehensive solution that meets your needs. For example, using a free lesson planning tool alongside a separate analytics platform can give you both the content creation and data insights you require. It's all about making the most of what's available. Look for tools that integrate well with each other and provide complementary features. Also, take advantage of any trial periods offered by paid tools. These trials can offer a glimpse into how a tool might fit into your teaching practice, helping you make informed decisions without immediate financial commitment.

Investing in paid AI tools requires a strategic approach. Start by identifying which features are most critical to your teaching. Are you looking for detailed analytics, personalized learning paths, or enhanced collaboration tools? Once you've pinpointed your priorities, research different options and compare what they offer. Budgeting for these tools involves considering the initial cost and any ongoing fees. It might be worthwhile to allocate funds from professional development or technology budgets to cover these expenses. In some cases, a premium tool offers the scalability and support that free alternatives lack, making it a valuable investment in the long run.

Success stories abound when it comes to using both free and paid AI tools in education. One educator shared how using a combination of free resources allowed her to create a tailored curriculum that met her students' diverse needs without straining her budget. Meanwhile, testimonials from those who have invested in premium services highlight the benefits of dedicated support and advanced features. They often note how these tools have enabled them to streamline administrative tasks, allowing for more personalized and engaging student interactions.

Ultimately, the choice between free and paid AI tools comes down to what best supports your teaching goals and enhances your classroom environment. By understanding the strengths and limitations of each option and employing a strategic approach, you can leverage AI to create a more dynamic and effective learning experience.

CURATED RESOURCE LIST OF AI TOOLS FOR STUDENT ENGAGEMENT

1. **Personalized and Adaptive Learning**: AI-driven learning systems that analyze student performance in real time and adjust content to fit individual needs. These platforms adapt to learning paces, offer additional exercises for struggling students, and accelerate content for advanced learners.

- **Knewton Alta**—An adaptive learning platform offering personalized educational content across various subjects, utilizing real-time analytics to support individual learning paths.
  - *Target Audience:* Secondary and college students.
  - *Cost:* Subscription-based.
  - *Link:* www.wiley.com/en-us/education/alta
- **DreamBox Learning**—An adaptive math and reading platform providing personalized lessons based on student performance, fostering conceptual understanding through interactive activities.
  - *Target Audience:* Elementary and middle school students.
  - *Cost:* Subscription-based with a free trial available.
  - *Link:* www.dreambox.com
- **MATHia**—An AI-driven math tutoring software offering personalized learning experiences, aligning with curriculum standards to support skill mastery.
  - *Target Audience:* Middle and high school students.
  - *Cost:* Subscription-based.
  - *Link:* www.carnegielearning.com/solutions/math/mathia/

- **eSpark Learning**—Provides adaptive instruction and practice in math, reading, and writing, offering personalized learning quests to engage students.
  - *Target Audience:* Students in grades K-8.
  - *Cost:* Free version available.
  - *Link:* www.esparklearning.com
- **Edmentum**—Offers adaptive curriculum and assessments to meet individual student needs, supporting personalized learning paths across various subjects.
  - *Target Audience:* K-12 students.
  - *Cost:* Subscription-based.
  - *Link:* www.edmentum.com

2. **AI-powered Tutoring and Chatbots**: AI-powered tutors and chatbots provide 24/7 academic assistance, helping students with questions, explanations, and study support.

- **Khan Academy's Khanmigo**—An AI-powered tutor and teaching assistant offering personalized guidance across subjects like math, humanities, and coding, engaging students in interactive learning experiences.
  - *Target Audience:* All grade levels.
  - *Cost:* Free for educators in over forty countries.
  - *Link:* www.khanmigo.ai
- **Socratic by Google**—Helps students find explanations and answers to questions via text, voice, or pictures, offering resources like videos, images, and text from sources like YouTube and Khan Academy.
  - *Target Audience:* Secondary and college levels.
  - *Cost:* Free to use.

- o *Link:* https://socratic.org
- **Microsoft Math Solver**—Offers solutions for a wide range of math problems, from basic arithmetic to advanced calculus, using AI to provide step-by-step guidance.
  - o *Target Audience:* All grade levels.
  - o *Cost:* Free to use.
  - o *Link:* https://mathsolver.microsoft.com
- **Mathspace**—Offers AI-driven, curriculum-aligned digital workbooks that give instant feedback and step-by-step guidance on math problems.
  - o *Target Audience:* Middle school through college.
  - o *Cost:* Subscription-based with a free trial. *Link:*
  - o https://mathspace.co
- **ChatGPT by OpenAI**—ChatGPT is a versatile AI chatbot capable of assisting students with a wide range of subjects. It can help answer questions, explain complex concepts, and provide study support, making it a valuable tool for personalized learning.
  - o *Target Audience:* Middle school through college.
  - o *Cost:* Free version available.
  - o *Link:* https://chat.openai.com
- **Other popular AI chatbots:**
  - o Microsoft Copilot: https://copilot.microsoft.com
  - o Claude: https://claude.ai
  - o Gemini: https://gemini.google.com
  - o Perplexity: www.perplexity.ai

3. **Student Engagement and Interactive Learning**: AI-driven tools that help bring interactive, multimedia-rich content, instant feedback, and gamified experiences to make learning an adventure.

- **Quizizz**—An interactive platform where teachers can create quizzes and polls with real-time feedback to enhance student engagement.
  - ○ *Target Audience:* All grade levels.
  - ○ *Cost:* Free version available.
  - ○ *Link:* https://quizizz.com
- **Kahoot!AI**—A game-based learning platform that allows educators to create quizzes and interactive lessons, fostering engagement through gamification.
  - ○ *Target Audience:* All grade levels.
  - ○ *Cost:* Free version available.
  - ○ *Link:* https://kahoot.com
- **Prodigy**—A game-based learning platform that uses AI to tailor questions to the student's grade level and proficiency, making math and English fun and engaging.
  - ○ *Target Audience:* Elementary and middle school levels.
  - ○ *Cost:* Free version available.
  - ○ *Link:* www.prodigygame.com
- **EdPuzzle**—Allows educators to import videos and embed quiz questions, voiceovers, and other interactive elements.
  - ○ *Target Audience:* All grade levels.
  - ○ *Cost:* Free version available.
  - ○ *Link:* www.edpuzzle.com
- **Minecraft: Education Edition**—Utilizes the popular game to create immersive learning experiences, teaching subjects like history, math, and coding.
  - ○ *Target Audience:* Elementary and secondary levels.
  - ○ *Cost:* Subscription-based with a free trial.
  - ○ *Link:* https://education.minecraft.net

- **Yousician**—AI-powered music learning app designed to teach users how to play various instruments, including guitar, piano, bass, ukulele, and vocals.
  - *Target Audience:* Suitable for all ages.
  - *Cost:* Free for a limited amount of lesson time each day. Subscription plans are available.
  - *Link:* https://yousician.com

4. **Immersive Learning with Virtual Reality and Simulations**: AI-powered VR and simulation tools that make learning interactive, allowing students to explore virtual environments.

- **Class VR**—A comprehensive virtual reality (VR) and augmented reality (AR) learning system that provides standalone VR headsets, teacher-controlled lesson plans, and interactive 3D educational content across various subjects. It is used in schools worldwide to create engaging, immersive learning environments for students.
  - *Target Audience:* Grades K to 12.
  - *Cost:* Subscription-based.
  - *Link:* www.ClassVR.com
- **Labster VR**—Virtual science labs for hands-on experiments covering biology, chemistry, and physics, enhancing STEM education through immersive simulations.
  - *Target Audience:* Secondary and higher education students.
  - *Cost:* Subscription-based.
  - *Link:* www.labster.com

- **PHET Interactive Simulations**—Developed by the University of Colorado Boulder, PHET provides interactive math and science simulations.
  - *Target Audience:* Secondary and higher education students.
  - *Cost:* Free.
  - *Link:* https://phet.colorado.edu
- **Eduverse by Avantis Education**—Immersive learning experiences across various subjects.
  - *Target Audience:* All grade levels.
  - *Cost:* Subscription-based.
  - *Link:* www.avantiseducation.com/eduverse

5. **Tools to Foster Creativity**: AI tools that empower students to become creators rather than mere consumers of information.

- **Book Creator**—An intuitive app that enables students and teachers to create interactive digital books. Users can combine text, images, audio, and video to craft engaging content that fosters creativity and engagement.
  - *Target Audience:* All grade levels.
  - *Cost:* Offers a free version. Subscription is available for additional functionalities.
  - *Link:* https://bookcreator.com
- **DALL-E**—An AI system capable of generating images from text descriptions. Developed by OpenAI, it is useful for creative projects in art and design.
  - *Target Audience:* Secondary and college levels.
  - *Cost:* It offers a free version with limited credits. Additional credits can be purchased. It is included in some versions of ChatGPT.

- *Link:* https://openai.com/index/dall-e-2
- **Blender**—Advanced 3D creation suite that supports modeling, animation, rendering, and more. Widely used for art, design, and game development projects.
  - *Target Audience:* Advanced high school and college levels.
  - *Cost:* Free.
  - *Link:* www.blender.org
- **Scratch**—A visual programming language developed by MIT, allowing kids to create games, animations, and interactive stories. AI-powered features guide learners through challenges.
  - *Target Audience:* Suitable for ages 7–16.
  - *Cost:* Free.
  - *Link:* https://scratch.mit.edu
- **Roblox Studio**—Uses Lua programming to let kids create their own games within the Roblox universe. AI-powered features guide students through scripting and debugging.
  - *Target Audience:* Suitable for ages 8–16.
  - *Cost:* Free version available.
  - *Link:* https://create.roblox.com

6. **AI-Driven Language Learning**: AI tools that enable real-time translation, pronunciation coaching, and adaptive language learning.

- **Duolingo AI**—Provides AI-powered language learning with interactive exercises and personalized feedback across numerous languages. Gamified lessons enhance engagement.
  - *Target Audience:* Suitable for all ages.

- o *Cost:* Free version available.
- o *Link:* www.duolingo.com
- **Rosetta Stone**—Utilizes AI to provide immersive language learning experiences with speech recognition technology aiding in pronunciation and conversation skills.
  - o *Target Audience:* Secondary and adult learners.
  - o *Cost:* Subscription-based with a free trial.
  - o *Link:* www.rosettastone.com
- **Lingvist**—An AI-driven language learning app that personalizes vocabulary and grammar lessons based on user performance.
  - o *Target Audience:* Secondary and college levels.
  - o *Cost:* Subscription-based with a free trial.
  - o *Link:* https://lingvist.com
- **Elsa Speak**—An AI-powered app focusing on English pronunciation, providing real-time feedback to help users improve their speaking skills. Ideal for non-native English speakers.
  - o *Target Audience:* Secondary and higher levels. It may be helpful in some cases for younger learners.
  - o *Cost:* Subscription-based with a free trial.
  - o *Link:* https://elsaspeak.com

7. **Reading, Writing, and Grammar Assistance**: AI tools that are designed to improve literacy and writing.

- **Lexia Core5 Reading**—An adaptive blended learning program that accelerates the development of literacy skills for students of all abilities.
  - ○ *Target Audience:* Elementary students.
  - ○ *Cost:* Subscription-based.

- *Link:* www.lexialearning.com/core5
- **Sora by OverDrive Education**—A student reading app providing access to ebooks and audiobooks from the school's library. Features include a built-in ebook reader, audiobook player, and tools for assignments. Sora is free for students to use, with schools managing their digital book collections.
  - *Target Audience:* All grade levels.
  - *Cost:* Free for students.
  - *Link:* www.overdrive.com
- **Grammarly**—Provides real-time grammar, spelling, and style suggestions to improve English writing quality.
  - *Target Audience:* Secondary and college levels.
  - *Cost:* Offers a free version.
  - *Link:* www.grammarly.com
- **Linguix**—An AI-based writing assistant offering grammar corrections, style improvements, and vocabulary enhancements in six languages.
  - *Target Audience:* Secondary and college levels.
  - *Cost:* Offers a free version.
  - *Link:* https://linguix.com
- **Hemingway Editor**—Analyzes text for readability, passive voice, and sentence complexity, helping writers create clear, concise content.
  - *Target Audience:* All grade levels.
  - *Cost:* Offers a free version.
  - *Link:* https://hemingwayapp.com

8. **Tools to Teach Coding**: AI tools designed to make programming fun, interactive, and accessible for kids of all ages. They help develop computational thinking,

problem-solving skills, and a strong foundation in coding.

- **Code.org**—AI-driven lessons and coding games that introduce kids to programming through engaging projects. It offers popular courses like Hour of Code and integrates with AI chatbots for assistance.
  - *Target Audience:* All grade levels.
  - *Cost:* Free.
  - *Link:* https://code.org
- **Tynker**—A game-based learning platform that uses AI-powered storytelling and puzzles to teach kids coding concepts in Python, JavaScript, and Scratch.
  - *Target Audience:* Suitable for ages 5–14.
  - *Cost:* Free version available.
  - *Link:* www.tynker.com
- **Kodable**—AI-driven drag-and-drop coding platform that introduces young learners to sequencing, loops, and conditionals through interactive games.
  - *Target Audience:* Suitable for ages 4–10.
  - *Cost:* Free version available.
  - *Link:* www.kodable.com
- **CodeMonkey**—Uses AI-driven challenges and puzzles to teach kids real programming languages (Python, CoffeeScript).
  - *Target Audience:* Suitable for ages 7–14.
  - *Cost:* Free trial available.
  - *Link:* www.codemonkey.com

9. **AI-Supported Special Education**: AI tools that improve accessibility by offering assistive technologies and personalized learning.

- **Microsoft Immersive Reader**—AI-powered tool for students with dyslexia and reading challenges. Designed to enhance reading comprehension through text-to-speech, visual aids, and customizable reading settings.
  - *Target Audience:* Suitable for all grade levels.
  - *Cost:* Free as part of Microsoft Office 365.
  - *Link:* www.microsoft.com/en-us/education/learning-tools
- **Otter.ai**—AI-generated real-time captions and transcriptions for students with hearing impairments.
  - *Target Audience:* Suitable for all grades
  - *Cost:* Free version available.
  - *Link:* http://otter.ai/education

- **Seeing AI (by Microsoft)**—Helps visually impaired students understand their surroundings by narrating the visual world. It can read text, describe scenes, and identify objects.
  - *Target Audience:* Suitable for all grade levels.
  - *Cost:* Free.
  - *Link:* Download on the App Store or Google Play.
- **Google Lookout**—AI tool for visually impaired students to identify objects and text. Free to use.
  - *Target Audience:* Suitable for all grade levels.
  - *Cost:* Free.
  - *Link:* Download for Android from Google Play.
- **Speechify**—AI-powered text-to-speech tool that assists students with reading disabilities by converting written text into spoken words.
  - *Target Audience:* Suitable for all grade levels.
  - *Cost:* Free version available.
  - *Link:* https://speechify.com

- **Proloquo2Go**—AAC (Augmentative and Alternative Communication) app for nonverbal students providing a voice through symbol-based communication.
  - *Target Audience:* Suitable for all grade levels.
  - *Cost:* Subscription-based.
  - *Link:* www.assistiveware.com/products/proloquo2go
- **Tobii Dynavox**—Assistive technology for communication and learning, designed for students with disabilities. Their products include eye-tracking devices and communication apps.
  - *Target Audience:* Suitable for all grade levels.
  - *Cost:* Subscription-based.
  - *Link:* https://us.tobiidynavox.com

10. **AI-Supported Homeschooling**: AI tools that provide innovative approaches to enhance the homeschooling learning process.

- **Time4Learning**—Online homeschool curriculum that utilizes adaptive learning technologies to personalize education. The platform covers subjects such as math, language arts, science, and social studies, providing interactive lessons and assessments.
  - *Target Audience:* Suitable for K-12.
  - *Cost:* Subscription-based.
  - *Link:* www.time4learning.com
- **Homeschool Panda**—Planning and management app for homeschooling families. It features lesson planning, record-keeping, and community support, integrating AI to personalize educational experiences.
  - *Target Audience:* Suitable for K-12.

      ○ *Cost:* Subscription-based.

      ○ *Link:* www.homeschoolpanda.com.

- **Outschool**—Online platform offering live classes across various subjects taught by qualified educators. It provides AI recommendations to match students with appropriate classes.
  - ○ *Target Audience:* Suitable for ages 3–18.
  - ○ *Cost:* Free and paid options are available.
  - ○ *Link:* https://outschool.com

- **Adventure Academy**—An immersive, game-based learning platform that uses AI to adapt lessons in math, reading, and science for young learners.
  - ○ *Target Audience:* Suitable for ages 8–13.
  - ○ *Cost:* Subscription-based with a free trial.
  - ○ *Link:* www.adventureacademy.com

11. **AI-Powered Student Well-Being and Mental Health Support**: AI tools that detect early signs of stress and provide emotional support tools for students.

- **Woebot**—AI chatbot providing mental health support through conversational interactions. It offers techniques for managing stress, anxiety, and depression.
  - ○ *Target Audience:* Suitable for ages thirteen and above.
  - ○ *Cost:* Free version available.
  - ○ *Link:* https://woebothealth.com

- **Wysa**—AI mental health app offering coping techniques and emotional support through chat-based interactions. It provides tools for managing stress, anxiety, and sleep issues.
  - ○ *Target Audience:* Suitable for ages thirteen and above.
  - ○ *Cost:* Free version available.

      o *Link:* https://wysa.com
- **BrainCo**—AI-driven focus and mindfulness tools designed to improve students' attention and emotional regulation. Its products include wearable devices and apps that provide real-time feedback.
  - o *Target Audience:* Suitable for K-12 students.
  - o *Cost:* Subscription-based.
  - o *Link:* www.brainco.tech/#/
- **Headspace AI**—AI-powered mindfulness and stress management tool offering guided meditations, sleep aids, and focus exercises.
  - o *Target Audience:* Suitable for K-12 students.
  - o *Cost:* Subscription-based.
  - o *Link:* www.headspace.com

CURATED RESOURCE LIST OF AI TOOLS FOR EDUCATORS

1. **Administrative and Educator Support Tools**: AI tools that automate administrative tasks like scheduling, grading, student enrollment, and communication.

- **Google Classroom**—Google Classroom integrates with various AI tools to automate administrative tasks, provide personalized feedback, and enhance collaboration between students and teachers. It streamlines assignment creation, distribution, and grading, facilitating efficient communication within the educational environment.
  - o *Target Audience:* All grade levels.
  - o *Cost:* Free for educators and students.
  - o *Link:* https://classroom.google.com

- **Ellucian**—AI-driven student information system designed for higher education institutions. It manages student data, enrollment, and academic records, providing analytics to support decision-making processes.
  - *Target Audience:* Universities and Colleges.
  - *Cost:* Subscription-based.
  - *Link:* https://www.ellucian.com

- **Fireflies.ai**—Ai-powered meeting assistant for recording and summarizing lectures.
  - *Target Audience:* Educators and administrators at all levels.
  - *Cost:* Free version available with limited features.
  - *Link:* https://beta.fireflies.ai
- **Google's Gemini for Education**—AI-powered co-teaching assistant.
  - *Target Audience:* All grade levels.
  - *Cost:* Free for educators.
  - *Link:* https://edu.google.com

2. **Assessment and Grading**: AI tools that automate grading, analyze student performance and provide detailed feedback beyond multiple-choice tests.

- **Gradescope**—An AI-assisted grading platform that streamlines the grading process for assignments and exams. It supports various assessment types, including problem sets and essays, providing detailed analytics on student performance.
  - *Target Audience:* Secondary and college levels.
  - *Cost:* Free version available with limited features.

      ○ *Link:* www.gradescope.com

- **Pear Deck Learning (Formerly Edulastic)**—Provides formative and summative assessment tools with AI-driven analytics to track student performance and identify learning gaps. It offers a vast library of standards-aligned assessments and real-time data to inform instruction.
  - *Target Audience:* Suitable for K-12.
  - *Cost:* Free version available.
  - *Link:* www.peardeck.com
- **TeachFX**—Uses AI to analyze classroom discussions and provide private feedback on student engagement and teacher talk time. This helps teachers improve their craft.
  - *Target Audience:* All grade levels.
  - *Cost:* Subscription-based.
  - *Link:* https://teachfx.com
- **PowerSchool Performance Matters**—Provides AI-driven analytics for student performance, helping educators identify trends and areas for intervention.
  - *Target Audience:* All grade levels
  - *Cost:* Subscription-based.
  - *Link:* www.powerschool.com

3. **Lesson Planning and Content Creation**: AI tools that help educators generate teaching materials such as lesson plans, quizzes, and interactive exercises.

- **MagicSchool AI**—Assists educators in generating lesson plans, assessments, and differentiated content tailored to diverse learning needs.
  - *Target Audience:* Suitable for all grade levels.
  - *Cost:* Free version available.

- *Link:* www.magicschool.ai
- **Brisk Teaching**—Provides AI tools for creating lesson plans, quizzes, and presentations.
  - *Target Audience:* Suitable for all grade levels.
  - *Cost:* Free for teachers.
  - *Link:* www.briskteaching.com
- **Curipod**—An AI-powered tool that assists teachers in creating interactive presentations and lessons tailored to student engagement.
  - *Target Audience:* Suitable for all grade levels.
  - *Cost:* Free version available.
  - *Link:* https://curipod.com
- **Education Copilot**—An AI-driven platform that helps educators design personalized lesson plans and interactive activities.
  - *Target Audience:* Suitable for all grade levels.
  - *Cost:* Free version available.
  - *Link:* https://educationcopilot.com
- **Canva for Education**—An AI-powered design tool offering a vast library of resources to quickly craft engaging slides, posters, worksheets, and videos. It's an intuitive platform for creating visually appealing educational materials. Free for primary and secondary educators worldwide.
  - *Target Audience:* Primary and secondary grade levels.
  - *Cost:* Free for educators worldwide.
  - *Link:* https://www.canva.com/education/

ENCOURAGING STUDENT FEEDBACK ON AI TOOLS

Students are not just passive recipients of AI-driven education but active contributors to its evolution. Their insights and experiences can significantly refine and improve the tools we use. When students share what works and what doesn't, it opens opportunities for meaningful improvements. This feedback loop ensures that AI tools remain relevant and effective, adapting to the ever-changing dynamics of a classroom. By listening to our students, we can create a more engaging and personalized learning environment that truly meets their needs.

There are practical techniques that can make this process straightforward and insightful. Anonymous surveys are an excellent starting point. They allow students to express their honest opinions without fear of judgment. These surveys can include questions about the usability of the AI tools, their impact on learning, and suggestions for improvement. Focus groups, on the other hand, provide a platform for students to engage in open discussions about their experiences. These groups can offer rich, qualitative insights that surveys might miss. Additionally, many AI tools feature real-time feedback options, enabling students to provide immediate input as they interact with the technology. This immediacy captures their genuine reactions and can be invaluable for making timely adjustments.

Responsive feedback systems play a crucial role in tailoring AI tools to student preferences. When students see that their feedback leads to tangible changes, it reinforces their value in the educational process. These systems can adjust settings or suggest features based on user input, creating a more intuitive and satisfying experience. For instance, if students find a particular AI feature confusing, feedback can prompt developers to simplify it or provide additional guidance. This adaptability improves the tool's effectiveness and fosters a sense of ownership among students. They feel heard and valued, knowing their input has a direct impact on their learning experience. This empowerment can increase engagement and motivation as students become active participants in shaping their education.

To further integrate student feedback into the educational process, consider using it to inform professional development. Educators can learn from student insights, identifying areas needing additional training or support. This growth mindset can enhance teaching practices and ensure educators are equipped to make the most of AI tools. Sharing feedback with colleagues can also foster a culture of collaboration and innovation as teachers work together to optimize their use of technology. By treating student feedback as a valuable resource, educators can continuously refine their approach, creating a learning environment that is both responsive and effective.

CHAPTER 10: PREPARING FOR FUTURE AI INNOVATIONS

As educators, staying ahead of emerging AI trends is crucial. Artificial intelligence is a transformative force reshaping how we teach and learn, and understanding these changes will empower you to harness its full potential in your classroom.

EMERGING AI TRENDS: WHAT EDUCATORS NEED TO KNOW

One of the most significant AI trends impacting education is predictive analytics. These systems analyze data from various sources to predict student performance and identify those needing additional support. By understanding patterns in student behavior, you can intervene early, providing targeted help and resources. This approach enhances learning outcomes and personalizes education, making it more relevant to each student's needs. Predictive analytics can be a game-changer in fostering an inclusive learning environment where every student can succeed. By leveraging these insights, you can tailor your teaching strategies to meet each student's unique needs, ultimately improving their academic journey.

Virtual and augmented reality are also making waves in education. These technologies offer immersive learning experiences that transport students beyond the confines of the traditional classroom. Imagine teaching biology with a virtual tour of the human body or history through a virtual walk-through of ancient Rome. These applications engage students and deepen their understanding of complex subjects. Virtual reality creates opportunities for experiential learning, allowing students to explore and interact with content in previously unimaginable ways. As these technologies become more accessible, they will play an increasingly important role in education, offering new ways to captivate and educate students.

Natural language processing is another key trend in revolutionizing educational tools. This technology allows computers to understand and respond to human language, facilitating student communication and AI-driven systems. Applications include AI-powered chatbots that provide instant feedback on assignments and language learning apps that offer real-time pronunciation guidance. Natural language processing can enhance the learning experience by providing students with immediate support and feedback. This technology empowers students to take control of their learning, fostering independence and confidence in their abilities.

AI-driven interactive whiteboards and smart displays are becoming staples in modern classrooms. These devices offer dynamic, interactive lessons, allowing students to engage with content like never before. Teachers can use these tools to create a more interactive and collaborative learning environment where students actively participate in their education. Smart displays can adapt to different learning styles, providing visual, auditory, and kinesthetic learners with tailored resources. This adaptability ensures that every student can engage with the material in a way that suits them best, enhancing overall comprehension and retention.

Advanced AI simulations are also emerging as powerful tools for experiential learning. These simulations allow students to apply their knowledge in real-world scenarios, bridging the gap between theory and practice. For example, chemistry students can conduct virtual experiments in a risk-free environment, while business students can run simulations to understand market dynamics. These experiences help students develop critical thinking and problem-solving skills, equipping them with the tools they need to succeed in the real world.

The implications of these trends for education are profound. As AI evolves, it will shape curriculum design, teaching methodologies, and classroom dynamics. The shift toward more student-centered learning environments is inevitable, as AI facilitates collaboration and personalized learning. By embracing these changes, educators can create engaging and inclusive classrooms where every student thrives. AI will not replace teachers; instead, it will augment their capabilities, allowing them to focus on what they do best: inspiring and educating the next generation.

However, integrating these innovations comes with challenges. Infrastructure upgrades are necessary to support new technologies, and educators must receive training to use advanced AI tools effectively. Schools must invest in professional development to ensure teachers have the skills and knowledge to integrate AI into their teaching practices. By providing educators with the resources and support they need, schools can overcome these challenges and unlock the full potential of AI in education.

To stay informed about these emerging AI trends, educators should seek out resources that provide insights and updates. AI trend reports and industry publications can offer valuable information about the latest developments in educational technology. Online communities and forums dedicated to AI in education provide a platform for educators to share experiences, ask questions, and collaborate on innovative solutions. By staying informed and connected, educators can ensure they are prepared to adapt to the ever-changing landscape of AI in education.

The classroom of tomorrow is just around the corner, but it's up to educators to seize the opportunities that AI presents and transform education for the better.

PREPARING STUDENTS FOR AN AI-DRIVEN WORLD

Imagine a world where understanding artificial intelligence is as important as knowing how to read or do basic math. This is the reality our students will face. To prepare them, we need to focus on essential AI skills beyond knowing how to use technology. They need to understand the principles behind AI, like how algorithms work and why data is so crucial. It's about developing critical thinking and problem-solving skills within an AI context. These skills help students analyze information, identify patterns, and make informed decisions. By fostering these abilities, we empower them to navigate an AI-driven world with confidence and creativity.

Incorporating AI into your curriculum doesn't mean overhauling everything. It's about weaving AI concepts into what you already teach. Start with the basics. Introduce AI through relatable examples in everyday life, like voice assistants or recommendation systems. As students become more comfortable, you can expand into more complex topics, such as machine learning and neural networks. Consider offering courses on AI ethics and its societal impacts. These classes can spark meaningful discussions about the role of AI in our lives, encouraging students to think critically about the technology they use daily. Project-based learning focused on AI applications can also be incredibly effective. Let students work on real-world problems using AI tools, giving them a hands-on experience that solidifies their understanding and makes learning engaging and practical.

To truly prepare students, we need to foster AI literacy from an early age. Think about developing programs that ignite curiosity and build foundational skills. AI clubs are a fantastic way to start. They allow students to explore AI topics, collaborate on projects, and share their discoveries. Extracurricular activities like coding competitions or robotics teams can also enhance AI literacy, allowing students to apply what they've learned in exciting, competitive settings. Workshops on AI tool usage and applications can further deepen their knowledge, teaching them how to leverage technology to solve problems and innovate. These programs not only build technical skills but also encourage creativity, teamwork, and resilience, all of which are crucial in an AI-driven world.

Promoting ethical AI awareness is just as important as teaching the technical aspects. Students must understand AI's ethical implications, such as bias, privacy, and accountability. Classroom debates on AI-related ethical dilemmas can stimulate critical thinking and help students develop their perspectives on these complex topics. You might present scenarios where AI decisions impact people's lives and ask students to discuss the pros and cons. These activities encourage them to consider the broader implications of AI and the responsibilities that come with using such powerful tools. Providing case studies that examine real-world AI challenges can also be enlightening. These stories offer insights into how AI is applied, the challenges it presents, and the solutions found. By engaging with these examples, students learn to appreciate the nuances of AI implementation and the importance of ethical decision-making.

Create a classroom environment that encourages open discussion about AI. Encourage students to ask questions and share their thoughts on how AI affects their daily lives. These conversations can lead to a deeper understanding and spark interest in further exploration. Consider inviting guest speakers who work in AI-related fields to share their experiences and insights. These interactions can provide students with a glimpse into potential career paths and the real-world applications of AI technology. By connecting classroom learning with real-life experiences, students can see the relevance of AI education and feel motivated to engage with the subject matter.

Incorporating AI education into your curriculum can also involve collaboration with other educators. Work with colleagues to develop interdisciplinary projects that integrate AI concepts across subjects. For example, a history teacher might collaborate with a computer science teacher to explore how AI can analyze historical texts and provide new perspectives on historical events. These collaborative efforts enrich the curriculum and demonstrate the interconnectedness of knowledge and the versatility of AI applications. By working together, educators can create a more cohesive and comprehensive AI education experience for their students.

In addition to curriculum integration, schools can support AI education by providing access to resources and technology. Ensure that students can access up-to-date software, hardware, and online platforms that facilitate AI learning. Schools can also collaborate with local businesses or organizations to secure funding or donations for AI-related resources. By providing students with the tools they need to explore AI, schools can create an environment where learning is accessible and engaging for all.

Ultimately, preparing students for an AI-driven world is about more than just teaching technical skills. It's about fostering a mindset of curiosity, adaptability, and ethical responsibility. Educators can shape the next generation of thinkers, innovators, and leaders, equipping them with the skills and knowledge they need to succeed in an AI-driven future.

AI AND CAREER READINESS: ALIGNING EDUCATION WITH FUTURE JOB MARKETS

Let's shift our focus to the world beyond the classroom, where AI is rapidly transforming the job market. As AI technologies become more sophisticated, they are reshaping what employers seek in potential hires. Roles that didn't exist a decade ago, like AI ethics officers and machine learning engineers, are now in high demand. The World Economic Forum highlights that AI, big data, and machine learning specialists are among the fastest-growing job categories. This trend underscores a critical need for students to acquire skills that align with these emerging roles. Today's learners must be prepared to use AI tools and innovate and drive AI development. This involves understanding complex algorithms, data analysis, and ethical considerations—skills that are becoming as fundamental as math and literacy.

As educators, aligning educational goals with these market demands is crucial. Curriculum development should reflect the changing landscape by incorporating AI-related skills into career and technical education. This shift requires rethinking what we teach and how we teach it. By integrating AI competencies, we can prepare students to thrive in industries that are increasingly reliant on technology. Collaborating with industry partners is an effective strategy to ensure our curricula meet real-world needs. These partnerships can provide insights into the skills and knowledge that companies value, helping educators design programs that equip students with relevant and practical expertise. By aligning educational goals with job market demands, we give students a competitive edge, empowering them to pursue lucrative and fulfilling careers in AI-driven fields.

Providing career guidance resources is another essential component of preparing students for the future. Career fairs focusing on AI offer students a glimpse into the diverse opportunities available in this field. These events can bring together industry professionals, educators, and students, fostering connections and offering insights into the various paths one can take with AI expertise. Workshops and mentorship programs further support this exploration. These programs provide valuable mentorship and guidance by connecting students with industry professionals. Students can gain first-hand knowledge about the skills needed to succeed in AI careers and the challenges and rewards of working in this dynamic field. Mentorship programs also offer students a support network, helping them confidently navigate their career paths.

Encouraging industry partnerships goes hand in hand with providing career guidance. Schools can foster collaborations with AI industries to create student internships and apprenticeship opportunities. These hands-on experiences are invaluable, offering students the chance to apply their learning in real-world settings. Internships and apprenticeships enhance students' technical skills and teach them to work effectively in team environments, communicate their ideas, and solve complex problems. Joint initiatives between schools and industries can also facilitate real-world AI project experiences. By working on projects that address actual industry challenges, students gain practical experience that prepares them for future employment. These partnerships benefit both students and industries, as companies gain fresh perspectives and potential future employees while students gain the experience, knowledge, and confidence needed to succeed.

Integrating AI into education is not just about preparing students for specific roles. It's about fostering a mindset of continuous learning and adaptability. As AI technologies evolve, so too will the skills required to work alongside them. By instilling a passion for lifelong learning, we prepare our students to navigate these changes with agility. Encourage students to take ownership of their learning, explore new technologies, and stay informed about industry trends. This proactive approach to education ensures that students remain relevant and competitive in a fast-paced job market.

In addition to technical skills, soft skills are equally important in the workplace. Creativity, critical thinking, and communication are vital assets for anyone looking to excel in AI-related fields. These skills enable individuals to approach problems from multiple angles, communicate complex ideas clearly, and work collaboratively with diverse teams. Educators can foster these skills by incorporating project-based learning, interdisciplinary collaborations, and opportunities for creative expression into their teaching practices. By nurturing both technical and soft skills, we equip students with a well-rounded education that prepares them for the demands of the modern workforce.

As we prepare students for careers in an AI-driven world, it's essential to recognize the broader societal implications of AI. Ethical considerations, such as data privacy, bias, and accountability, must be at the forefront of AI education. Encourage students to engage with these issues critically, understanding the impact of their work on society. By promoting ethical AI practices, we ensure that the next generation of AI professionals is equipped to develop technologies that benefit humanity as a whole. Educators play a crucial role in shaping this future, guiding students as they become responsible and conscientious contributors to the AI landscape.

In closing, the intersection of AI and career readiness is a pivotal focus for educators. By aligning educational goals with job market needs, providing robust career guidance, and fostering industry partnerships, we can empower students to thrive in an AI-driven world. The skills and experiences they gain in the classroom today will prepare them to innovate, lead, and make a positive impact in their chosen fields. As we continue exploring AI's role in education, let's keep our eyes on the horizon, preparing students to participate in the future and shape it.

CONCLUSION

As you close this book, I hope you feel both inspired and empowered by the possibilities AI brings to education. Together, we've explored how AI can transform classrooms—personalizing learning, boosting student engagement, and making education more accessible. But AI is more than just a tool; it's an opportunity to reimagine teaching in a way that keeps students at the heart of learning.

The goal has never been to replace the human touch that makes teaching so powerful. Instead, AI should serve as a partner, amplifying your ability to connect with students, support diverse learning needs, and create dynamic, inclusive classrooms. By integrating AI thoughtfully and ethically, you can enhance—not diminish—the human side of education.

Throughout this book, we've shared practical strategies for integrating AI into your classrooms. From interactive AI tools to predictive analytics, the examples we've explored demonstrate how AI, when used responsibly, can drive meaningful change. But this is just the beginning. AI in education is not a one-time shift—it's an ongoing journey of learning, adapting, and refining best practices.

Continuous learning is crucial. As AI evolves, so should your professional development. Stay informed about emerging trends and technologies and be open to new methods and ideas. This mindset will help you adapt and thrive in the ever-changing landscape.

So, where do you go from here? **Start small.** Experiment with AI-powered tools that align with your teaching goals. **Stay curious.** Continue learning about emerging AI trends and how they can enhance your classroom experience. **Collaborate.** Share insights with fellow educators, involve parents and students, and build a community of support and innovation.

Most importantly, approach AI with a critical yet open mind. As an educator, you are uniquely positioned to ensure AI is used in ways that uplift students and create opportunities for all.

The future of education is being shaped right now, and you are a part of that transformation. By embracing AI with intention and purpose, you're not only preparing students for a world where AI plays an integral role—you're leading the way.

MAKE A DIFFERENCE WITH YOUR REVIEW

Your Words Can Help Shape the Future of AI in Education

People who share their experiences inspire and guide others. Your review could be the encouragement another educator needs to embrace AI in their classroom!

Would you help a fellow educator—curious about AI but unsure where to start?

My mission is to make **The Complete Guide to AI in Education** accessible and practical for all educators. But to reach more teachers, I need your help.

Most people choose books based on reviews. By sharing your thoughts, you can help:

- One more teacher feel confident using AI tools.
- One more student benefit from personalized learning.
- One more school integrate AI ethically and effectively.
- One more classroom become more engaging and inclusive.

It takes just a minute, costs nothing, and could make all the difference.

To leave a review, simply **scan the QR code below**:

Thank you from the bottom of my heart!

Gloria Lembo

REFERENCES

"5 AI Concepts Defining the Future of Work in 2024." Factnewsph, February 8, 2024. https://factnewsph.org/2024/02/5-ai-concepts-defining-the-future-of-work-in-2024.html.

"Balancing Tech and Life: How Students Use AI to Enhance Daily Living." 1883 Magazine, January 31, 2023. https://1883magazine.com/balancing-tech-and-life-how-students-use-ai-to-enhance-daily-living/.

"Inspiring Students to Dream & Empowering Them to Achieve." *Connections Magazine* (blog), August 23, 2023. https://www.connectionsmag.com/elementary-school-districts/antioch-school-district-34/inspiring-students-to-dream-empowering-them-to-achieve/.

"Optimizing Schools." Princeton, n.d. https://aiethics.princeton.edu/wp-content/uploads/sites/587/2018/10/Princeton-AI-Ethics-Case-Study-3.pdf.

"Research-Based Educational Strategies." Class Intercom, June 8, 2023. https://classintercom.com/research-based-educational-strategies/.

"What Are the Pros and Cons of AI for Image Design?" Rumie, n.d. https://learn.rumie.org/jR/bytes/what-are-the-pros-and-cons-of-ai-for-image-design/.

98thPercentile. "Creative AI Coding Classes for Kids," n.d. https://www.98thpercentile.com/artificial-intelligence-coding.

AccessForce. "Explore Accessible Universities," n.d. https://accessforce.org/schools/accessibility/davenport-university-warren-location/.

Admin. "Data Science Archives." *Technology Point* (blog), n.d. https://technologypoint.in/category/data-science/.

Admin-science. "Artificial Intelligence in Education 2023 - Unleashing the Power of Technology in Schools," January 23, 2024. https://mmcalumni.ca/blog/artificial-intelligence-revolutionizing-education-in-2023-transforming-learning-with-advanced-technology.

Admin-science. "Utilizing ChatGPT for Educators: Enhancing Teaching and Learning Experience." Study in Artificial Intelligence in Canada, January 23, 2024. https://mmcalumni.ca/blog/gpt-3-a-revolutionary-tool-for-educators-to-enhance-teaching-and-learning.

AI Arsenal. "7 Different Types of Artificial Intelligence," August 19, 2023. https://ai47labs.com/blogs/7-different-types-of-artificial-intelligence/.

AI for Education. "AI Course for Educators: Ready to Unlock the Power of AI in Your Teaching Practice?," n.d. https://www.aiforeducation.io/ai-course.

Akgun, Selin, and Christine Greenhow. "Artificial Intelligence in Education: Addressing Ethical Challenges in K-12 Settings." *Ai and Ethics* 2, no. 3 (2022): 431–40. https://doi.org/10.1007/s43681-021-00096-7.

Akvelon. "Empowering Your Business With Local LLMs Becomes Possible," n.d. https://akvelon.com/empowering-your-business-with-local-llms-becomes-possible/.

Ali. "AI Tutors: Unlocking Potential with ChatGPT." vRealm, October 29, 2023. https://www.thevrealm.com/post/ai-tutors-chatgpt-guide.

Alshiha, Mada Bandar, and Ahlam Mohammed Al-Abdullatif. "Gamification in Flipped Classrooms for Sustainable Digital Education: The Influence of Competitive and Cooperative Gamification on Learning Outcomes." *Sustainability* 16, no. 23 (December 6, 2024): 10734. https://doi.org/10.3390/su162310734.

Aqua. "Discover the Transformative Potential of AI in Education." *AI News* (blog), April 19, 2023. https://aquariusai.ca/blog/discover-the-transformative-potential-of-ai-in-education.

Arjun C Vinod, A Ananthakrishnan, A.R Abhishek, S. Adithyan, and Tintu Varghese. "Is Artificial Intelligence a Threat or a Benefit?" *International Journal of Engineering Technology and Management Sciences* 6, no. 5 (September 28, 2022): 553–55. https://doi.org/10.46647/ijetms.2022.v06i05.087.

AscentionDX. "Where Can I Get Some?" June 27, 2023. https://www.ascensiondx.com/the-power-of-artificial-intelligence.

AVID Open Access. "AI and the 4 Cs: Critical Thinking," n.d. https://avidopenaccess.org/resource/ai-and-the-4-cs-critical-thinking/.

b12admin. "The Future of Education: How Digital Transformation Is Shaping Learning." *B12 App* (blog), January 8, 2024. https://www.b12app.com/the-future-of-education-how-digital-transformation-is-shaping-learning/.

Badia, Amit. "AI and Me: Understanding Artificial Intelligence." Ab Infocom, October 11, 2023. https://www.abinfocom.com/blog/artificial-intelligence-13/ai-and-me-17.

Baraishuk, Dmitry. "What Is Artificial Intelligence? AI vs Traditional Software." Belitsoft, January 21, 2024. https://belitsoft.com/ai-development/artificial-intelligence-vs-conventional-software.

Biederwolf, Kyle. "Comcast's Internet Essentials Program Connected More Than 480,000 Low-Income Coloradans Over the Past 10 Years." Comcast Colorado, April 14, 2021. https://colorado.comcast.com/2021/04/14/comcasts-internet-essentials-program-connected-more-than-480000-low-income-coloradans-over-the-past-10-years/.

CDW.com. "AI for Education: Best Strategies for Integrating AI With Learning Objectives," September 18, 2024. https://www.cdw.com/content/cdw/en/arti

cles/software/ai-education-best-strategies-integrating-ai-learning-objec tives.html.

Code.org. "AI 101 for Teachers," n.d. https://code.org/ai/pl/101.

Cook, Jessica, and Steve Baule. "Leveraging AI to Help Special Education Teach- ers." eSchool News, August 2, 2024. https://www.eschoolnews.com/digital- learning/2024/08/02/leveraging-ai-to-help-special-education-teachers/.

Cornell University. "Ethical AI for Teaching and Learning," n.d. https://teaching.- cornell.edu/generative-artificial-intelligence/ethical-ai-teaching-and- learning.

Cornell University. "Ethical AI for Teaching and Learning," n.d. https://teaching. cornell.edu/generative-artificial-intelligence/ethical-ai-teaching-and- learning.

CoSN. "CoSN Presents 10 Essential Back-to-School Resources for a Successful Academic Year," n.d. https://www.cosn.org/cosn-news/cosn-presents-10- essential-back-to-school-resources-for-a-successful-academic-year/.

Course Hero. "Thomas Jefferson High School for Science and Technology," n.d. https://www.coursehero.com/sitemap/schools/150835-Thomas-Jefferson- High-School-for-Science-and-Technology/.

Dhingra, Sifatkaur, Manmeet Singh, Vaisakh S.B., Neetiraj Malviya, and Sukhpal Singh Gill. "Mind Meets Machine: Unravelling GPT-4's Cognitive Psychology." *BenchCouncil Transactions on Benchmarks, Standards and Evaluations* 3, no. 3 (September 2023): 100139. https://doi.org/10.1016/j.tbench.2023.100139.

Diaz, Carlos. "Different Approaches to Online Math Tutoring." *Urban Splatter* (blog), June 30, 2023. https://www.urbansplatter.com/2023/06/different- approaches-to-online-math-tutoring/.

DigitalDefynd Team. "Use of AI in Schools [25 Case Studies] [2025]." *DigitalDefynd* (blog), May 8, 2024. https://digitaldefynd.com/IQ/ai-in-schools-case-studies/.

Disco. "Top 5 AI Tools for Collaborative Learning in 2025," December 18, 2024. https://www.disco.co/blog/ai-tools-for-collaborative-learning.

Ditch That Textbook. "40 AI Tools for Teachers, Educators and Classrooms (Free and Paid)," November 13, 2023. https://ditchthattextbook.com/ai-tools/.

Editorial Team. "Comparing Traditional Pedagogical Methods with AI-Driven Adaptive Learning Systems." Thekeyfact, January 3, 2024. https://thekeyfact. com/comparing-traditional-pedagogical-methods-with-ai-driven-adaptive- learning-systems/.

EdTech Update. "Top EdTech Update Elementary Assessment Content for March 2024," March 2024. https://www.edtechupdate.com/?edition=monthly&date= 2024-03.

Ellwood, Mark. "The Impact of AI on Job Roles and Skills." Ellwood Consulting, n.d. https://www.ellwoodconsulting.com/blog/2024/07/the-impact-of-ai-on-job-roles-and-skills.

Ersozlu, Zara, Sona Taheri, and Inge Koch. "A Review of Machine Learning Methods Used for Educational Data." *Education and Information Technologies* 29, no. 16 (November 1, 2024): 22125–45. https://doi.org/10.1007/s10639-024-12704-0.

Express, Windhoek, and Bjorn Wiedow. "A Case for Early Technology Education." *Windhoek Express*, July 12, 2024. https://www.we.com.na/technology-we/a-case-for-early-technology-education2024-07-12.

Fitzpatrick, Dan. "Philadelphia Schools To Embrace AI In 2025." Forbes, December 18, 2024. https://www.forbes.com/sites/danfitzpatrick/2024/12/18/philadelphia-schools-to-embrace-ai-in-2025/.

Foran-Mulcahy, Katie. "Research Guides: AI Tools for Education: Curipod." University of Cincinnati, n.d. https://guides.libraries.uc.edu/ai-education/cu.

Forward, Squad, and Educational Tools. "Teachology AI Review: Best AI Lesson Plan Generator." *Educational Tools* (blog), September 24, 2024. https://educational.tools/teachology-ai-review-best-ai-lesson-plan-generator/.

Galileo. "AI-Powered Digital Classrooms: The Future of Education," n.d. https://www.higalileo.ai/blog/ai-powered-digital-classrooms-the-future-of-education.

Galileo. "AI-Powered Digital Classrooms: The Future of Education," n.d. https://www.higalileo.ai/blog/ai-powered-digital-classrooms-the-future-of-education.

Giboney, Sara. "How to Use Artificial Intelligence as a Study Tool." Nebraska Methodist College, January 20, 2023. https://blog.methodistcollege.edu/how-to-use-artificial-intelligence-as-a-study-tool.

Gobir, Nimah. "8 Free AI-Powered Tools ThQEDat Can Save Teachers Time and Enhance Instruction." KQED, October 3, 2023. https://www.kqed.org/mindshift/62462/8-free-ai-powered-tools-that-can-save-teachers-time-and-enhance-instruction.

Gonzales, Jose, and Yogini Joglekar. "How AI and VR Enhance the Immersive Learning Experience." *Training Industry* (blog), May 7, 2024. https://trainingindustry.com/articles/learning-technologies/how-ai-and-vr-enhance-the-immersive-learning-experience/.

Gupta, Chandan. "How AI Is Shaping the Future of STEM Education in Rural America." MRCC EdTech, October 14, 2024. https://mrccedtech.com/ai-is-transforming-stem-education/.

Hi-Tech Weirdo. "Where Did the Roblox Game Phenomenon Come From?" November 26, 2023. https://hitechweirdo.com/where-did-the-roblox-game-phenomenon-come-from/.

Iracst. "How to Use TikTok to Teach Tech Skills and Boost Digital Literacy," n.d. https://www.iracst.org/how-to-use-tiktok-to-teach-tech-skills-and-boost-digital-literacy.

Jackson, Elle. "Student Engagement Strategies for Teachers in 2024 and Beyond." Digital Joy, October 19, 2023. https://www.digitaljoy.media/student-engagement-strategies-for-teachers-in-2024/.

Jayaraj, U. "Machine Learning." Certisured, n.d. https://certisured.com/quickypedia/machine-learning.

Klein, Alyson. "AI's Potential for Bias Puts Onus on Educators, Developers." GovTech, June 27, 2024. https://www.govtech.com/education/k-12/ais-potential-for-bias-puts-onus-on-educators-developers.

Langreo, Lauraine. "No, AI Won't Destroy Education. But We Should Be Skeptical." *Education Week*, August 31, 2023, sec. Technology, Artificial Intelligence. https://www.edweek.org/technology/no-ai-wont-destroy-education-but-we-should-be-skeptical/2023/08.

Leopold, Till. "Future of Jobs Report 2025: The Jobs of the Future – and the Skills You Need to Get Them." World Economic Forum, January 8, 2025. https://www.weforum.org/stories/2025/01/future-of-jobs-report-2025-jobs-of-the-future-and-the-skills-you-need-to-get-them/

Marcinek, Andrew. "Engaging Parents in the AI Conversation." *Medium* (blog), March 15, 2024. https://medium.com/@andrewmarcinek/engaging-parents-in-the-ai-conversation-4c6277631d87.

Marino, M. "Harnessing Flipped Learning to Enhance Student Engagement: A Transformative Approach." Education Articles, July 3, 2023. http://education-articles.com/harnessing-flipped-learning-to-enhance-student-engagement-a-transformative-approach/.

Mohamed, Amr M., Tahany S. Shaaban, Sameh H. Bakry, Francisco D. Guillén-Gámez, and Artur Strzelecki. "Empowering the Faculty of Education Students: Applying AI's Potential for Motivating and Enhancing Learning." *Innovative Higher Education*, October 7, 2024. https://doi.org/10.1007/s10755-024-09747-z.

Nagelhout, Ryan. "Better Feedback with AI?" Harvard Graduate School of Education, November 17, 2023. https://www.gse.harvard.edu/ideas/usable-knowledge/23/11/better-feedback-ai.

Nguyen, Andy, Ha Ngan Ngo, Yvonne Hong, Belle Dang, and Bich-Phuong Thi Nguyen. "Ethical Principles for Artificial Intelligence in Education." *Education and Information Technologies* 28, no. 4 (April 1, 2023): 4221–41. https://doi.org/10.1007/s10639-022-11316-w.

NHA Communications Team. "Windemere Park Charter Academy Wins Building the Hope Award for Outstanding Academic Achievements." Windemere Park

Charter Academy, January 11, 2024. https://www.nhaschools.com/schools/windemere-park-charter-academy/en/blog/windemere-park-wins-building-the-hope-award.

Office of Communications. "Thinking Like A Computer | Thomas Jefferson High School for Science and Technology." FCPS, September 29, 2023. https://tjhsst.fcps.edu/features/thinking-computer.

Ouyang, Fan, and Liyin Zhang. "AI-Driven Learning Analytics Applications and Tools in Computer-Supported Collaborative Learning: A Systematic Review." *Educational Research Review* 44 (August 1, 2024): 100616. https://doi.org/10.1016/j.edurev.2024.100616.

Pedagog.ai. "Learn More About Us," January 17, 2023. https://pedagog.ai/about/.

Pizza Tower. "Geometry Dash Scratch | Play Online Now," October 2, 2023. https://pizzatower.io/geometry-dash-scratch/.

Poth, Rachelle Dené. "7 AI Tools That Help Teachers Work More Efficiently." Edutopia, October 20, 2023. https://www.edutopia.org/article/7-ai-tools-that-help-teachers-work-more-efficiently/.

Practice Benefit Corp. "Embracing AI in Education: A Cost-Effective Revolution for Personalized Learning," n.d. https://practicebc.com/insights/embracing-ai-in-education-a-cost-effective-revolution-for-personalized-learning/.

Schools & Community Psychology Service. "Support with Exam Anxiety." Smart School Services, July 11, 2024. https://psychology.smartschool.services/supporting-children-young-people-with-exam-anxiety/.

Schools That Lead. "10 Best AI Tools to Become a More Efficient Teacher," September 4, 2024. https://www.schoolsthatlead.org/blog/best-ai-tools-to-become-efficient-teacher.

Sharma, Ravi. "Future of Education: AI Compliance with FERPA and GDPR - Hurix Digital." Digital Engineering & Technology | Elearning Solutions | Digital Content Solutions, October 30, 2024. https://www.hurix.com/blogs/data-privacy-in-education-through-ferpa-and-gdpr-adherence/.

Socrates, Astakhov. "Who Uses Speech Synthesis?" Antivivisection information about AI and VR, February 7, 2024. https://antivivisection.info/who-uses-speech-synthesis/.

Ta, Linh. "Some Ankeny Teachers Are Bringing AI into Classrooms." Axios, August 23, 2024. https://www.axios.com/local/des-moines/2024/08/23/ai-classrooms-ankeny-des-moines-schools.

Taskiner, Murat. "Celebrating Transformation: Success Stories of AI in Education." LinkedIn, February 2, 2024. https://www.linkedin.com/pulse/celebrating-transformation-success-stories-ai-murat-taskiner-feo0e.

TechLeadership. "Yousician," n.d. https://www.techleadership.ch/technology-1/yousician.

The Brand Evolution. "Maximizing Brand Communication: How AI Is Revolution-izing Efficiency," March 4, 2025. https://the-brandevolution.com/ai-brand-communication/.

Thomas, Jilson. "What Are the Pros and Cons of AI for Image Design?" Rumie, n.d. https://learn.rumie.org/jR/bytes/what-are-the-pros-and-cons-of-ai-for-image-design/.

TJREVERB. "TJ Space," n.d. https://activities.tjhsst.edu/cubesat/.

Tolson, Courtney. "Positive Classroom Culture." Kreative Science, February 22, 2023. https://www.kreativescience.com/post/positive-classroom-culture.

Turn.io. "Unleashing the Power of Chatbots in Education: Insights and Strategies," July 12, 2023. https://www.turn.io/news/unleashing-the-power-of-chatbots-in-education-insights-and-strategies.

UNESCO. "Recommendation on the Ethics of Artificial Intelligence," n.d. https://www.unesco.org/en/articles/recommendation-ethics-artificial-intelligence.

University at Buffalo. "AI + Education Learning Community Series," n.d. https://ed.buffalo.edu/ai-ed.

Verma, Nikita. "How Effective Is AI in Education? 10 Case Studies and Examples." *Axon Park* (blog), February 8, 2023. https://axonpark.com/how-effective-is-ai-in-education-10-case-studies-and-examples/.

Vhearts. "Join Online Coding Classes For Kids - LogicLearning," n.d. https://vhearts.net/post/287000_empower-your-child-039-s-future-with-fun-and-interactive-online-coding-classes-f.html.

Xu, Weiqi, and Fan Ouyang. "The Application of AI Technologies in STEM Educa-tion: A Systematic Review from 2011 to 2021." *International Journal of STEM Education* 9, no. 1 (September 19, 2022): 59. https://doi.org/10.1186/s40594-022-00377-5.

ZGM.org. "The Impact of Virtual Reality on Education: Enhancing Learning in the Digital Age," August 22, 2023. https://zgm.org/2023/08/22/impact-of-vr-on-education/.

Printed in Dunstable, United Kingdom

67070123R00097